CAMBRIDGE LIBRARY COLLECTION

Books of enduring scholarly value

Technology

The focus of this series is engineering, broadly construed. It covers technological innovation from a range of periods and cultures, but centres on the technological achievements of the industrial era in the West, particularly in the nineteenth century, as understood by their contemporaries. Infrastructure is one major focus, covering the building of railways and canals, bridges and tunnels, land drainage, the laying of submarine cables, and the construction of docks and lighthouses. Other key topics include developments in industrial and manufacturing fields such as mining technology, the production of iron and steel, the use of steam power, and chemical processes such as photography and textile dyes.

Rivers and Canals

Leveson Francis Vernon-Harcourt (1839–1907) drew on a distinguished career in canal and river engineering for this illustrated two-volume survey, here reissued in its enlarged 1896 second edition. Having started as an assistant to the civil engineer John Hawkshaw, Vernon-Harcourt was appointed resident engineer in 1866 for new works on London's East and West India docks. Later, as a consulting engineer, he specialised in the design and construction of harbours, docks, canals and river works, and he was elected professor of civil engineering at University College London in 1882. This publication covers the design and construction of tidal and flood defences, canals, locks, and irrigation works. Volume 1 covers the physical characteristics of rivers and estuaries, and the control of their flow through dredging and works such as weirs and breakwaters. Vernon-Harcourt also discusses the design of flood defences. His *Harbours and Docks* (1885) is also reissued in this series.

Cambridge University Press has long been a pioneer in the reissuing of out-of-print titles from its own backlist, producing digital reprints of books that are still sought after by scholars and students but could not be reprinted economically using traditional technology. The Cambridge Library Collection extends this activity to a wider range of books which are still of importance to researchers and professionals, either for the source material they contain, or as landmarks in the history of their academic discipline.

Drawing from the world-renowned collections in the Cambridge University Library and other partner libraries, and guided by the advice of experts in each subject area, Cambridge University Press is using state-of-the-art scanning machines in its own Printing House to capture the content of each book selected for inclusion. The files are processed to give a consistently clear, crisp image, and the books finished to the high quality standard for which the Press is recognised around the world. The latest print-on-demand technology ensures that the books will remain available indefinitely, and that orders for single or multiple copies can quickly be supplied.

The Cambridge Library Collection brings back to life books of enduring scholarly value (including out-of-copyright works originally issued by other publishers) across a wide range of disciplines in the humanities and social sciences and in science and technology.

Rivers and Canals

*With Statistics of the Traffic
on Inland Waterways*

VOLUME 1: RIVERS

LEVESON FRANCIS VERNON-HARCOURT

CAMBRIDGE
UNIVERSITY PRESS

CAMBRIDGE
UNIVERSITY PRESS

University Printing House, Cambridge, CB2 8BS, United Kingdom

Cambridge University Press is part of the University of Cambridge.
It furthers the University's mission by disseminating knowledge in the pursuit of
education, learning and research at the highest international levels of excellence.

www.cambridge.org
Information on this title: www.cambridge.org/9781108080590

© in this compilation Cambridge University Press 2015

This edition first published 1896
This digitally printed version 2015

ISBN 978-1-108-08059-0 Paperback

RIVERS AND CANALS

VERNON-HARCOURT

VOL. I.

𝕷ondon

HENRY FROWDE

OXFORD UNIVERSITY PRESS WAREHOUSE
AMEN CORNER, E.C.

𝕹ew 𝕐ork

MACMILLAN & CO., 66 FIFTH AVENUE

RIVERS AND CANALS

THE
FLOW, CONTROL, AND IMPROVEMENT OF RIVERS

AND THE

DESIGN, CONSTRUCTION, AND DEVELOPMENT OF CANALS
BOTH FOR NAVIGATION AND IRRIGATION

*WITH STATISTICS OF THE TRAFFIC
ON INLAND WATERWAYS*

BY

LEVESON FRANCIS VERNON-HARCOURT, M.A.

MEMBER OF THE INSTITUTION OF CIVIL ENGINEERS
AUTHOR OF 'HARBOURS AND DOCKS,' AND 'ACHIEVEMENTS IN ENGINEERING'

IN **TWO** VOLUMES.—VOL. I

RIVERS

SECOND EDITION, RE-WRITTEN AND ENLARGED

Oxford

AT THE CLARENDON PRESS

1896

𝔒𝔵𝔣𝔬𝔯𝔡

PRINTED AT THE CLARENDON PRESS

BY HORACE HART, PRINTER TO THE UNIVERSITY

PREFACE

—•—

WHEN, early in 1892, my publishers informed me
that a second edition of ' Rivers and Canals would be
soon required, two alternative courses appeared to
be open to me. Motives of expediency indicated the
advisability of merely revising the book, so as to bring
it up to date ; for in the first edition of a book, with
costly illustrations, and of which more than half the
copies have been sold in America and foreign countries,
at the usual reduced rate, the author's share of the
profits little more than suffices to defray the expenses
incurred in the preparation of the drawings for the
plates and woodcuts. On the other hand, the experience gained, since the book was first written, from
attending six international navigation congresses, in
visiting numerous rivers and canals on the continent,
by inspecting professionally a number of rivers in the
United Kingdom, in writing ten papers on river and
canal works for the Institution of Civil Engineers,
international congresses, and other societies, and in
the preparation of full reports upon the condition and
improvement of several rivers, could not have been
embodied in a mere revise of the book. Under these

circumstances, it appeared incumbent on me to dismiss all considerations of personal advantage, and to endeavour to the best of my ability, and with such opportunities as my scanty leisure time afforded, to utilize widened experience, increased knowledge, and the additional published matter available, in the preparation of as complete a book as practicable, within reasonable limits, on river and canal engineering.

Accordingly, with the exception of some paragraphs from the first edition, incorporated here and there in the earlier part of the book, and especially in the first two chapters, the book has been practically re-written. Rivers and canals have been dealt with in two separate parts, forming respectively the two volumes, as distinct from one another as the intimate connection of the two subjects will allow, instead of being mixed up together as in the previous book; and the various branches of each subject have been classified, as far as possible, under separate chapters. Although several of the plans and sections illustrating the previous edition have necessarily been reproduced, but with additions and to a reduced scale, the plates have been entirely remodelled; and a considerable majority of the drawings did not appear at all in the former set of plates. Some of the least important of the previous illustrations have been omitted; but the plans have been so reduced and arranged, that a considerably larger number of illustrations have been brought together within a smaller space than formerly; similar scales have been adopted, as far as possible, for the plans and sections in the same plate; and the colouring introduced into the drawings, at the suggestion of the lithographer, adds greatly to their clearness.

Besides re-writing the descriptions of river and canal works given in the first edition, under the light of more recent experience and fuller information, numerous other works have been described, such as for instance the training works of the Loire, the Weser, the Nervion, and the Ribble, irrigation canals, the inland navigation systems of various countries with their traffic, and the Manchester, Baltic, Corinth, and Nicaragua canals. The principal publications from which information has been derived are given in the footnotes, to render reference to them easy; and plans and sections, kindly furnished me by engineers, have been acknowledged in a similar manner. I also desire to express my sense of the assistance I have derived, in the preparation of this book, from the excellent technical library of the Institution of Civil Engineers, the published Proceedings of that Institution, and the Papers read at the international inland navigation congresses held within the last ten years.

My chief aims in this book have been to endeavour, by the investigation of the various physical conditions affecting rivers, as illustrated by observations, the results of works, and experimental inquiries, to place the principles of river engineering upon a more scientific basis, and to give, for the benefit of engineers, concise accounts of the chief works carried out for the control and improvement of rivers, and the construction of canals. Elementary considerations and explanations have been retained in the earlier chapters of the book, with the object of rendering it useful to engineering students; whilst it is hoped that the descriptions of important works of general interest, with the aid of the illustrations, will render the book acceptable to the public, and that statistics relating to traffic on inland waterways and ship-canals will prove of interest to traders.

In conclusion, I would fain venture to hope that my earnest endeavours to render this book worthy of the important branch of engineering with which it deals, a means of promoting the progress of river engineering on scientific lines, and valuable generally to engineers, will meet with the approval of my professional brethren at home and abroad; and if this book should succeed in gaining their approbation, I shall regard it as my highest and most enduring reward.

L. F. VERNON-HARCOURT.

6 QUEEN ANNE'S GATE, WESTMINSTER.
December 9, 1895.

P.S. The final revision of a portion of the text and the completion of the index, have had to be accomplished, under somewhat difficult conditions, during a voyage to India, taken for the purpose of inspecting the River Hooghly, and reporting thereon to the Commissioners of the Calcutta Port Trust.

CONTENTS OF VOL. I

—◆—

CHAPTER I.

PHYSICAL CHARACTERISTICS.

CHAPTER II.

MEASUREMENT AND FORMULÆ OF DISCHARGE.

CHAPTER III.

REGULATION AND CANALIZATION OF RIVERS.

CHAPTER IV.

DREDGING AND EXCAVATING.

CHAPTER V.

LOCKS; AND FIXED, AND DRAW-DOOR WEIRS.

CHAPTER VI.

MOVABLE WEIRS.

CHAPTER VII.

PREDICTION OF FLOODS ; AND PROTECTION FROM INUNDATIONS.

CHAPTER VIII.

DELTAS OF TIDELESS RIVERS ; AND IMPROVEMENT OF THEIR OUTLETS.

CHAPTER IX.

JETTIES AND BREAKWATERS AT THE MOUTHS OF RIVERS.

CHAPTER X.

TIDAL FLOW IN RIVERS; AND FORMS OF ESTUARIES.

CHAPTER XI.

DREDGING IN TIDAL RIVERS.

CHAPTER XII.

TRAINING WORKS IN ESTUARIES.

CHAPTER XIII.

TRAINING WORKS IN ESTUARIES (*continued*).

CHAPTER XIV.

EXPERIMENTAL INVESTIGATIONS ON TRAINING WORKS IN ESTUARIES.

LIST OF ILLUSTRATIONS IN VOL. I

—•—

RIVERS.

PLATE I.

HYDROLOGY OF RIVERS.

PLATE II.

DREDGING AND EXCAVATING MACHINES.

PLATE III.

REGULATION WORKS: FIXED AND DRAW-DOOR WEIRS.

REGULATION WORKS.

LIST OF WOODCUTS IN VOL. I

— ✦ —

ERRATUM.

Page 128, line 16 from top, *for* **Moskowa** *read* Moskva.

RIVERS AND CANALS.

PART I. RIVERS.

CHAPTER I.

PHYSICAL CHARACTERISTICS.

Introduction. River Basins. Differences of Rainfall. Variations in Rainfall. Evaporation and Percolation. Influences of Forests and Vegetation. Effect of Period of Rainfall. Impermeable and Permeable Strata. Available Rainfall. Forms of River Valleys. Fall of Rivers. Transportation of Material. Influences of Lakes. Divergence of Current. Tidal, and Tideless River Outlets. Importance of Scientific Study of Rivers.

RIVERS form a natural and easy means of communication between the sea and the interior of a country, and afford safe and convenient roadsteads for vessels. They also furnish the chief sources of water-supply; and the most fertile districts are situated along their banks. Consequently most of the important cities of the world have been built on the banks of rivers. Rivers, however, are not always suitable for navigation, in their natural condition, even in the lower portions of their course; and, owing to the continual changes taking place in their channels and at their outlets, they are liable to deteriorate if left to themselves. Moreover, rivers, whilst serving as the main arteries for the drainage of a country, and proving most valuable for irrigating lands in hot countries during the dry season, and as a source of water-power, are liable to devastate their valleys by extensive floods during periods of excessive rainfall. Accordingly, the regulation, improvement, and control of rivers constitute one of the most

B

important, and at the same time one of the most difficult
branches of civil engineering.

The natural state of any river depends upon a variety of
physical conditions, such as the area of its basin ; the amount,
distribution, and period of the rainfall over this drainage area ;
the nature of the superficial strata of its basin, and the extent
to which the surface is covered by vegetation ; the slopes of
the valleys of the main river and its tributaries ; the inclination
of the river-bed, and the material of which it is composed ;
and the amount of detritus brought down by the current.
On approaching the sea, a river is, in addition, affected by
the tidal rise in the sea at its outlet, by the size and form
of its estuary, by the sediment introduced by the flood tide,
by the direction of the prevailing winds, the travel of material
along the coast and the heaping-up action of the waves,
and by the tendency of the material brought down by the
river to deposit at its mouth in tideless seas.

River Basins.   The basin of a river is the whole area drained
by the river, or the area over which the rainfall tends to flow
into the river or its tributaries.   The boundary line of a
river basin is the ridge along the summit of the valleys of
the river and its tributaries, on each side of which the rainfall
flows in opposite directions into two different river basins,
just as the rain runs into different gutters on either side of
the ridge of a roof.   This boundary line between adjacent
basins is termed the water-parting, from the separation of
the flow on each side of it.

The size of a river, beyond the reach of tidal influences,
is naturally in some measure proportionate to the area which
it drains ; whilst the area of a river basin is dependent on the
extent of the continent, and its general configuration.   Large
river basins are only found in extensive continents, where
plains stretch for long distances inland from the coast.   The
insular position, and small extent of Great Britain preclude
it from possessing large rivers ; and even the river basins of

the European continent appear somewhat insignificant when compared with those of the vast continents of Asia, Africa, and America.

Amongst English rivers, the Thames comes first with a basin of 5,244 square miles. Next in importance is the Severn, with a basin of 4,350 square miles, which is followed by the Yorkshire Ouse, with a basin of 4,133 square miles; whereas the Dee, the Mersey, and the Ribble, though possessing large tidal estuaries, drain areas of only 813, 1,724, and 815 square miles respectively. The Tyne, the Tees, and the Clyde, also, though affording deep-water access to flourishing seaports, owing to the improvements effected in their tidal channels, have basins of merely 1,130, 735, and 945 square miles.

Turning to the principal rivers of Europe, draining far more extensive tracts of lands, the Seine has a basin of 30,370 square miles, about six times the drainage area of the Thames (Plate 1, Fig. 1); the Rhone has a basin of about 38,000 square miles; and the Loire, a basin of 45,000 square miles; whilst the lengths of the main stream of these rivers exceed 400 miles. Proceeding further into the interior of Europe, considerably larger river basins are met with ; thus, for instance, the Elbe has a drainage area of 55,000 square miles, and a length of 700 miles; the Rhine has a basin of 75,000 square miles, and a length of 800 miles ; the Don and the Dneiper have basins of 170,000, and 245,000 square miles respectively, with lengths of about 1,300 miles; the Danube has a basin of 312,000 square miles, with a length of 1,700 miles; and the Volga has a basin of 563,000 square miles, with a length of about 2,000 miles.

Considerable uncertainty exists as to the exact drainage areas of the rivers in the larger, and less explored continents ; but there are several which undoubtedly exceed the largest river basins of Europe. Thus in Asia, whilst the Ganges, with a basin of about 470,000 square miles and a length of about 1,500 miles, and the Yang-tse-kiang, with a basin of about 620,000 square miles and a length of 3,200 miles, as well as the Hoang-ho or Yellow River and the Amur, equal or exceed the largest rivers of Europe, they are surpassed by the Yenisei with a basin estimated at 880,000 square miles and a length of 3,700 miles, and the Lena draining central Siberia, and having a basin of 940,000 square miles. Moreover the largest river of Asia, the Obi, flowing through western Siberia, and falling into the Arctic Ocean after a course of 3,200 miles, has a drainage area of about 1.300,000 square miles, more than double the extent of the largest European river basin.

In Africa, the Nile is generally reputed to have the largest basin, with an area estimated at 1,500,000 square miles, the length of the river being supposed to exceed 4,000 miles. The next largest river basins are the Congo, with an area of 1,350,000 square miles, the Niger, 1,150,000 square miles, the Zambesi, 850,000 square miles, and the Orange River, 400,000 square miles, the last of which alone comes within the limits of the largest river basin of Europe.

South America derives its large river basins from its peculiar configuration, with the lofty chain of the Andes rising close along its western coast, and vast flat plains stretching eastwards; so that its rivers which flow into the Atlantic Ocean nearly traverse the continent, having their sources in the perennial snows of the western mountain range. Accordingly, the Amazon, flowing across the broadest part of the continent, draining an area of about 2,250,000 square miles, and having a length of nearly 4,000 miles, possesses the largest basin in the world; whilst the La Plata basin, with an estimated area of 1,600,000 square miles, exceeds in extent any river basin in the eastern hemisphere. The Mississippi is by far the largest river of North America, having a basin of 1,244,000 square miles, four times that of the Danube, and a length of 4,200 miles; whilst the

St. Lawrence has a basin about equal in extent to the Danube basin.

From the foregoing statistics, it will be seen that drainage areas may vary from the small gullies of streams rising close to the sea-coast—like Blackgang and Shanklin chines, in the Isle of Wight—up to the vast basins of mighty rivers, extending over great portions of the largest continents of the globe.

Differences of Rainfall. As rivers originate from rainfall, and form the channels by which the surplus rainfall flowing off the land is conveyed to the sea, observations concerning the rainfall over the basin of a river are very valuable in studying the régime of any river. Generally the rainfall is greatest in mountainous districts, especially where the hills rise near the coast, and where the prevalent winds blow from the ocean. These warm winds arriving fully laden with the moisture collected in their passage over the ocean, are arrested by the high land, and having their temperature reduced in rising up the mountain slopes, deposit their moisture as rain in passing over the mountain range. On the contrary, rainless districts are found in the interior of large continents, cut off from the sea by intervening mountain chains which deprive the air of its moisture before it reaches these regions. Accordingly, very great differences in rainfall are experienced in various parts of the world, ranging from zero over the deserts of the centre of Asia and Africa, up to an average annual rainfall of 474 inches at Cherrapunji in Assam, on the Khasi Hills, 4,455 feet above sea-level [1], the station having by far the heaviest recorded rainfall in the world.

Sometimes the amount of rainfall varies greatly in different parts of the same country. Thus whilst India possesses in its north-eastern corner the rainiest station known, and Uttray Mullay in Southern India has an average annual rainfall of 263 inches, the hill- station of Leh to the north

---

[1] 'The Climates and Weather of India, Ceylon, and Burmah.' H. F. Blanford.

of the Himalayas has an average rainfall of only 2 7 inches, and Jacobabad in Central India an average fall of only 4·4 inches in the year. Mexico and Peru also have very rainy, and rainless, districts only moderate distances apart. Even in England, the average annual rainfall ranges between 22 inches at Ely, and 175 inches at The Stye in Cumberland. Considerable differences in the rainfall may also be met with within the limits of a single moderate-sized river basin, varying with the elevation and with the proximity to the sea. Thus the rainfall over the Seine basin, which is a fairly level basin of moderate extent, ranges from under 24 inches up to over 48 inches in a year (Plate 1, Fig. 1); whilst in the Severn basin, which extends into the mountainous districts of North Wales, the average annual rainfall ranges from under 30 inches up to over 80 inches. Accordingly, the various portions of a river basin may be subject to very different meteorological influences, affecting the flow of the tributaries draining them; and the flow of the main river cohsists of a combination of these various influences, which are successively imparted to it at the confluence of its several tributaries. A plan, therefore, of a river basin, showing the main river and its tributaries, and also indicating the different average amounts of rainfall over this area, enables 'some idea to be formed of the comparative influence of the rainfall on the various tributaries, and indicates the places at which the flow of the main river is thereby affected (Plate 1, Fig. 1).

Variations in Rainfall. The amount of rainfall in the same locality varies from year-to year; and a reliable average can only be obtained from observations extending over several years. In some years, the rainfall is much below the average, leading to a deficiency of water for navigation in the dry season ; in other years, the rainfall is unusually large, causing the rivers to overflow their banks and inundate large tracts of land. Thus, in the exceptionally dry year 1887, the rainfall over the British Isles was only two-thirds of the average; whilst in the very wet year 1872, the rainfall exceeded the average by 36 per cent., and in 1877 by 25 per cent.[1] The rainfall of very wet years exhibits generally a greater deviation from the average than in very dry years ; and the proportionate variation of the rainfall is greater in dry districts than in wet ones. Taking the driest and wettest stations in England where records are kept, the rainfall was 14·63 inches at Ely, and 130·90 inches at The Stye in 1887, and 27·20 inches, and 243·98 inches respectively in 1872. The maxima and minima, however, do not occur at every station in the wettest and driest years, for the rainfall at Ely was 29·03 in 1877, or nearly 2 inches more than in 1872 ; whilst some stations experienced more rainfall in 1887 than in 1884, the excess amounting to nearly 2 inches at the very dry station of Hunstanton.

The yearly fluctuations of rainfall are greater in tropical countries than in the temperate zones, especially in places where the rainfall is moderate ; for whereas the rainfall of 1872 only attained about three times the rainfall of 1887 in two or three places in England, the maximum rainfall in India and Burmah amounts in many parts to three or four times the minimum rainfall, and even considerably more at a few places.

The rainfall, moreover, varies at different seasons of the year ; and the periods during which the maxima and minima occur depend on the locality, for the rainfall is due to the prevalence of the special winds blowing from the ocean, to which the locality may be most exposed. There are generally a wet and a dry season, extending over well-defined periods, within the tropics, and in places visited by periodical winds such as the monsoons ; but in temperate climates, where the winds are more variable, there is not a similar regularity in the rainy periods. Nevertheless, extended observations indicate that in temperate regions, districts

[1] 'British Rainfall, 1887.' G. J. Symons.

possess, on the average, a special period of maximum rain-fall, differing according to the period at which the wettest winds generally prevail, and the position of the district in relation to the ocean. Even in the British Isles, the ordinary period of maximum rainfall varies according to the situation; for it occurs in December and January on the extreme western coasts, it is in August in the eastern midlands of England, whilst over the rest of Great Britain, October is the wettest month. There are similar differences also in the periods of minimum rainfall.

Evaporation and Percolation. The rain falling on a river basin does not all find its way into the main river which drains the basin and discharges into the sea. Some of the rainfall is drawn up again into the air by the evaporation produced by the sun and wind; a portion sinks into the soil and affords moisture to trees and plants; and a portion fills subterranean cavities which form the sources of springs.

Evaporation varies with the locality, the season of the year, the state of the weather and the wind, and the nature of the soil on which the rain falls: it is greater from a surface of water than from the ground in summer, and from the ground in winter. The effects of evaporation over extensive areas of water are illustrated most forcibly by the state of the Caspian Sea, which, though it receives the waters of the Volga and other large rivers, and has no out-let, has had its water-level gradually reduced by the excess of evaporation over the combined discharges of these rivers, so that it is now about 85 feet below the level of the Black Sea. The waters of the Jordan also fail to raise the level of the Dead Sea, which lies in a great depression with its water-level 1,300 feet below the Mediterranean; whilst the discharge from the 750,000 square miles draining into Lake Chad is entirely absorbed by evaporation.

Evaporation attains its maximum during hot, dry weather, immediately succeeding rain on flat impervious ground, or

from a sheet of water; and it is least in cold, damp
weather.

Percolation depends on the nature of the soil, and the
weather; it is small in amount through earth, but large
through sand; and no percolation takes place in warm summer weather. The greatest amount of percolation occurs
with melting snow, and especially with small falls of snow
followed by a thaw. Water penetrating into the ground
to a depth of two or three feet appears to be removed from
the influence of evaporation.

Influences of Forests and Vegetation. A covering of trees
and grass is very beneficial in increasing the average discharge of a river by reducing evaporation, especially in the
summer months when water is most needed, and in equalizing
its flow during the rainy season, especially in the steeper
parts of the basin, by retarding the arrival of the rain to
the watercourses.

The clearing of forests tends to reduce the rainfall of a
district; but the main cause why forests and grass increase
the average flow of rivers is their arrest of evaporation, in
some instances the evaporation in the open country having
been found five times as great as the evaporation under
trees. The shelter afforded by the foliage prevents some
of the rain from reaching the ground, but this is much more
than compensated for by the smallness of the evaporation;
so that the effective rainfall in a forest is considerably
greater than in the open country, independently of the
larger rainfall over a forest, the proportion of which would
be difficult to determine. Forests and vegetation are
specially valuable on impermeable strata and on mountain sides, where, by regulating the downward flow of the
rain, and by binding together the loose soil, they diminish
the rapid rise of torrential streams, arrest the washing away
of the soil, and prevent the formation of gullets which
facilitate the descent of the water. Extensive clearings of

forests, and close cropping of pastures have led to such injurious denudation and floods in mountainous regions, that it has been found necessary to enforce the replanting of trees and the promotion of the growth of vegetation, which have produced beneficial results.

Effect of Period of Rainfall. The total amount of rain falling over the area of a river basin, in any given period, is not necessarily any measure of the discharge of the river. The rainfall over the Seine basin is greater in summer than in winter; but a glance at the diagrams showing the daily heights of the Seine from 1878 to 1890 (Plate 1, Figs. 3 to 6), indicates that large floods never occur in the river between May and October, whilst they are very frequent in the other half of the year.

As evaporation is very active in the summer, most of the rain which falls during the summer months is returned to the atmosphere, and only a small proportion finds its way into the rivers, unless the slope on which it falls is very steep, the strata of the basin impermeable, and the rainfall very great in a short period. Winter rains, on the contrary, find their way almost entirely into the river which drains the basin on which they descend, as they fall upon ground saturated by the wet of autumn, and when evaporation is nearly suspended. A continuous, or very heavy rainfall produces a greater effect on the discharge of a river than the same amount of rain distributed over a longer period. In many cases the greatest river floods are produced when a heavy rainfall takes place on melting snow.

As summer rains, in temperate regions, produce much less effect on the flow of rivers than rains in the winter and spring, a drought in the early autumn may often be predicted at the commencement of the summer, if the rainfall has been small in the preceding winter and spring.

In considering the flow of rivers within the temperate

zones, the year may be conveniently divided into the warm
and cold seasons, extending, in the northern hemisphere,
from May to October, and from November to April respectively, floods being generally only prevalent during the
cold season. (Plate 1, Figs. 3 to 6.)   An exception, however,
in the period of the occurrence of floods, must be noted in
the case of rivers originating from glaciers and snow-clad
mountain heights ; for the melting of the ice and snow by
the summer heat produces floods in these rivers, whereas
the cold of winter reduces their flow to a minimum.   The
natural divisions of the year in respect of rivers in tropical
countries are the dry and rainy seasons ; for the rivers are
nearly dried up towards the close of the first period, and
are immoderately swollen by the periodical rains.

Impermeable and Permeable Strata.   The nature of the
strata forming the basin of a river and its tributaries exerts
a very marked influence on its flow.

Where the strata are impermeable, the rain falling over
the basin is rapidly discharged by numerous little rivulets
into the main stream, and the river is torrential in its character.   It rises very rapidly in rainy weather, and its fall
is also rapid ; and in continued dry weather it becomes more
or less dried up.   As, however, no strata are absolutely impermeable, some of the rain sinks into the fissures in the
ground, and is delayed in its flow to the river.   Accordingly,
the rise of a torrential river is always more rapid than its
fall, which is delayed by the later arrival of the rain that
has penetrated below the surface.   These characteristics of
torrential rivers are illustrated by the flood diagrams of the
Brenne, the Yonne, and the Marne (Plate 1, Figs. 8, 9, and
11), tributaries of the Seine in the steeper, impermeable,
and rainiest portion of its basin.

Where the strata are permeable, the rain sinks into the
ground, and only finds its way gradually into the main
stream.   Consequently the rise of a river traversing per-

meable strata is more regular than that of a torrential river. The floods, however, of rivers flowing over permeable strata are more injurious to the adjacent lands, owing to their longer continuance, than the more rapidly rising floods of torrential streams which quickly subside. The flood diagrams of the Little Seine and the Aisne (Plate 1, Figs. 7 and 14), exhibit those properties of gently-flowing rivers draining permeable strata, and present a marked contrast to torrential floods.

The basins of large rivers generally comprise several kinds of strata, possessing different degrees of permeability, so that some of the tributaries receive chiefly the drainage of impermeable strata, and others of permeable soils, of which the tributaries of the Seine just referred to are instances (Plate 1, Fig. 2). The principal river possesses, in such cases, mixed characteristics, which depend on the proportion of permeable and impermeable strata over the basin, and may vary in different parts of its course, and even at different periods according as the rainfall predominates over one or other portion of the basin. This result is exemplified by comparing the rise of the same flood in the Seine above and below the confluence of the Yonne, and of the Marne, where the influence of the floods of these rivers on the floods of the Seine is clearly manifested (Plate 1, Figs. 7, 10, and 13). An advantage of a diversity in the strata of the tributary basins consists in the ordinary absence of coincidence in the arrival of the floods from the several tributaries at any point of the main river, owing to the rapid propagation and subsidence of torrential floods, which often have passed off before the floods from the gently-flowing tributaries have reached their maximum. Accordingly, an exceptional flood is more likely to result from a casual coincidence in the arrival of the floods from the various tributaries, than merely from a heavy rainfall over a basin comprising various strata.

Available Rainfall. If a relation could be established

between the volume of water which finds its way into a river, and the rainfall over the river basin, for strata of different permeability, it would be only necessary to ascertain the area of the basin, and the average nature of the superficial strata, in order to deduce the discharge of the river for any given rainfall. This proportion is specially affected by the degree of permeability of the strata forming the basin; for whereas the rain falling on an impermeable stratum finds its way rapidly into the river, when falling on a permeable stratum it sinks into the soil, and besides affording a considerable amount of moisture to vegetation, it supplies springs which either do not flow into the river at all, or in many cases join it at lower parts of its course.

The discharge of the River Po, draining an impermeable basin, has been found to amount to 75 per cent. of the rainfall; whilst the discharge of the Garonne above Marmande, and of streams draining clay beds, has been estimated at 65 per cent. of the rainfall. In the case of the Saône, with a less impervious basin, the discharge is equivalent to half of the rainfall. In the Upper Seine basin, however, where the permeable strata predominate (Plate 1, Fig. 2), the discharge does not quite amount to a quarter of the rainfall; whereas in the almost wholly permeable basin of the Eure, the discharge is only 15½ per cent. of the rainfall. Accordingly, it appears from these observations, that the available rainfall, or the proportion of the rainfall which actually reaches a river, may vary from 75 per cent. on impermeable strata, down to 15 per cent. on very permeable soils. These proportions, however, are merely yearly averages, and would be much less in the warm season, and greater in the cold season.

Forms of River Valleys. As a river forms its own bed in the bottom of the valley through which it flows, the nature of the bed depends on the geological strata which lie on the surface of the valley, or which the river has laid bare in the course of ages. The general form of the bottom of a river valley varies according as the strata are impermeable or permeable. The water flowing down the slopes of valleys composed of impermeable strata gradually wears away the projections, and fills up the hollows with soil from above, so that the slopes become curved as indicated in the sketch (Fig. 1). The rain, however, falling on

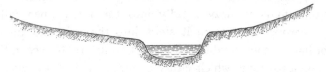

Section of the Valley of a River on Impermeable Strata.

Fig. 1.

a permeable river basin, as it sinks into the soil, does not materially modify the form of the slopes. Accordingly the valley, in this case, generally retains its original form, being nearly level at the base, or sometimes slightly raised along the banks of the river in consequence of the successive deposits produced by the river in flood-time, as shown on the accompanying sketch (Fig. 2).

Section of the Valley of a River on Permeable Strata.

Fig. 2.

Fall of Rivers. The fall of the bed of a river corresponds approximately to the general fall of the land over which it flows, and varies considerably in different localities. Usually the fall is greatest near the source of the river, gradually diminishing as it approaches the sea (Fig. 3); but this reduction in fall is not always continuous, for sometimes the fall increases again after the river has traversed a flat

table-land, or passed through a lake, as indicated by the
diagram.   If a river valley is terminated by a high mountain

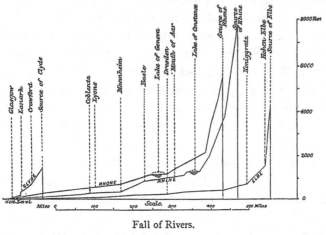

Fall of Rivers.

Fig. 3.

range, the river commences as a torrent, with a more or
less rapid fall according to the declivity of the mountain
side. If the mountains rise near the sea into which the
river flows, the average fall of the river is considerable, as
on the western side of America.   When, on the contrary,
the mountain ranges are separated from the ocean by vast
plains, as on the eastern side of the Rocky Mountains and
the Andes, the general fall of the river is gentle on emerging
from the mountains.

Transportation of Material. Rivers, besides conveying
rainfall to the sea, are also the transporters of alluvium.
Gradually, though slowly, all rivers are bringing detritus
down their valleys; and the materials of the mountains are
being carried to the sea. Glaciers, in their slow but irresistible
descent, grind the rocks of the mountain slopes down which
they imperceptibly glide, and when melting in the valleys
carry a large quantity of matter along with them. The
slopes of the hills are also disintegrated by frost and rain,

and the débris are brought down by the watercourses in flood-time. The velocity of the flow of rivers in the earlier portion of their course enables them to hold this matter in suspension, or to roll it along their bed; but as the fall of their bed, and consequently their velocity decreases, the heavier particles fall to the bottom, or come to rest, and tend to fill up the river-bed. The larger detritus is gradually reduced to gravel, sand, and silt by long-continued attrition, and is carried down the river in successive stages by floods, or is strewn over the valley. This deposit of material specially occurs when the velocity of the stream is abruptly checked by a sudden reduction in fall, as for instance where the River Po, on leaving the slopes of the Alps, descends into the plains of Lombardy, or when the flow of a river is arrested on entering a tideless sea.

The power of a current to transport material varies with the velocity of the current, and depends upon the specific gravity of the material, and the size of the particles. The results of a variety of experiments indicate that a current with a velocity of 3 to 6 inches per second will transport silt and clay, of 8 inches per second will transport sand, and of about 1 to 2½ feet per second will carry along gravel; whilst a current with a velocity of 6 feet per second will roll stones along. As the current slackens, the materials are sorted, the heaviest particles being deposited first.

Influences of Lakes. Lakes exercise a beneficial influence on rivers which flow through them, by arresting the alluvium brought down by the rivers, and by regulating their flow. The current of a river is checked on encountering the inert waters of the lake, which leads to the gradual deposit in the bottom of the lake of the sediment with which the river is charged. Thus the Rhone, on entering the lake of Geneva, deposits the whole mass of detritus, which it has brought down from the Alps, near the head of the lake, and issues from the lake at Geneva as pure and clear as the waters of the

lake ; and it presents a remarkable contrast to the turbid
Arve, which comes straight down from the Alps, and with
which its waters are mingled about a mile only below Geneva.
A river in flood flowing into a lake produces only a very slight
rise over the whole area of the lake, and is thus prevented
from rapidly swelling the discharge of the river below, which
has accordingly a much more uniform flow than it had
above the lake. Thus the Rhone and the Rhine are more
regular in their flow on issuing from the lakes of Geneva
and Constance, than in the upper portions of their course ;
and the river St. Lawrence, being fed almost entirely by
the lakes Superior, Huron, Michigan, Erie, and Ontario, is
so uniform in its flow that its level remains nearly always
constant.

Divergence of Current. The instability of a river channel
is not confined to the bottom of its bed. A river rarely flows,
especially across a flat, alluvial plain, in a straight and
uniform channel. A very slight impediment, such as a fallen
tree or a hard projection at one bank, or any irregularity
in the bed, will direct the main current against the opposite
bank, which, if composed of soft materials, is gradually
eroded ; and deposit collects near the other bank, so that
the course of the river is by degrees modified. After im-
pinging against the concave part of one bank, the current
is turned again at the next bend to the other bank, thus
gradually producing and intensifying the serpentine course
so noticeable in many rivers when flowing through plains
(Fig. 4, p. 18). The greatest current runs close along the
concave bank, which, besides promoting the erosion of the
bank, produces also a scour at the bottom. Accordingly,
in winding rivers, the channel is deepest close to the concave
bank, and gradually shoals towards the convex bank, as
shown in the section ; so that the deepest part of the channel
shifts across the river between two successive bends. There
is always deep water along the concave bank, as the current,

in being turned round the bend, necessarily keeps close to this bank ; whereas the current in tending across the river at the change in curvature of the bank, is less confined, and more variable in direction, so that the channel is shallower between the bends (Plate 3, Fig. 1).

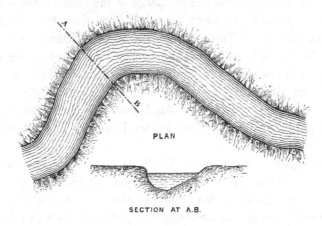

PLAN

SECTION AT A.B.

Fig. 4.

River Outlets. The mouths of rivers exhibit considerable diversities, depending on the form, size, and general physical conditions of the estuary which connects them with the sea, and the rise of the tide in the sea into which they flow. A broad distinction must be drawn between rivers up which the tide flows and ebbs for some distance, and the size of whose outlets depends mainly on the tidal ebb and flow, and rivers discharging into tideless seas, whose outlets, like their channels above, are wholly due to their fresh-water discharge. Tidal rivers generally emerge into large estuaries or bays before reaching the sea, affording large spaces for the influx of the tide, as for instance the Thames, the Mersey, the Humber (Plate 7, Figs. 7, 9, and 11), the Severn, the Seine (Plate 9, Fig. 1), the Garonne, the Scheldt, the Elbe, the Weser (Plate 9, Fig. 8), and the St. Lawrence. Tideless rivers, on the contrary, form continually advancing deltas by the

deposit of their alluvium in front of their outlets, on entering
the denser inert sea ; and they meander through the delta
they have heaped up in the sea, in several encumbered and
constantly lengthening channels, of which class of rivers the
Rhone, the Po, the Danube, the Volga, the Nile, and the
Mississippi are notable examples (Plate 5, Figs. 1, 2, and 5).
In tidal rivers, the sedimentary matter, which is partly carried
down from inland, and partly brought in by the flood tide
from the erosion of neighbouring coasts and sandbanks, is
kept in almost constant motion under ordinary conditions,
and is carried out to sea by the ebb tide ; whereas the
material accumulating at the mouths of tideless rivers is
entirely composed of detritus carried down by the river, and
is only partially removed by littoral currents and wave disturbance. The bar at the mouth of a tidal river is formed
by the heaping-up action of the sea on the beach, and the
travel of sand and shingle along the coast under the influence
of the prevailing winds, tending to close up the mouth of the
river ; but this barrier is to some extent lowered by the ebb
and flow of the tide, aided by the fresh-water discharge. The
bar in front of the outlet of a tideless river is due to the
deposit of alluvium, resulting from the slackening of the river
current on emerging from its channel into the sea.

Intermediate between these two distinct classes are some
rivers which, owing to the flatness of the alluvial land they
traverse on approaching the sea, and the small tidal rise at
their mouth, have formed several outlet channels, of which
the mouths of the Rhine and the Maas are instances ; and
there are others which, in consequence of their large flood
discharge, and the enormous quantities of sediment they bring
down, overpower the tidal flow, and form a delta, such as the
Ganges, the Irrawaddi, and the Orinoco.

Scientific Study of Rivers. The diversified physical conditions of rivers, as indicated by the foregoing considerations,
have frequently led engineers to express the opinion that each

river must be considered quite independently, without reference to the experience gained on other rivers. Physicians might with equal reason have declared that, as men are so different in constitution, the treatment of each person should be considered independently, in which case little progress would have been effected in medical science. In investigating any particular river, it is undoubtedly an engineer's duty to examine, as far as practicable, the various physical characteristics of the river ; but an intelligent investigation of any river assists in determining the special influences of the several physical conditions, and thus every fresh river studied affords further insight into the bearing and effect of each special condition. The relative importance of the several physical conditions may, indeed, vary in every river; but the natural laws which regulate the influence of each condition remain unaltered ; and the effects produced are the combined results of all the influences, in their several proportions, in each special case

Probably no river basin has been studied so carefully as that of the Seine; and the records furnished by Mr. Belgrand and his successors [1] have enabled me to draw up a fairly complete summary of the physical characteristics and hydrology of this river (Plate 1), which may form a model of the information desirable with respect to other river basins.   Such details, together with a longitudinal section and cross-sections of the river, charts of the outlet at different periods, and tidal diagrams, furnish the engineer with valuable material in designing improvement works.   The results of these works, moreover, should be carefully followed, so that they may all form onward steps in the science of river engineering.

[1] 'La Seine, Études Hydrologiques,' E. Belgrand ; and ' Service Hydrométrique du Bassin de la Seine.'

CHAPTER II.

MEASUREMENT AND FORMULÆ OF DISCHARGE.

Importance of Discharge Measurements. Definition of Terms:—Discharge; Slope; Sectional Area; Wetted Perimeter; Hydraulic Mean Depth; Mean Velocity. Measurement of Cross-Sections. Velocity Observations:—Floats; Current-Meters; Gauge-Tubes; Hydrodynamometer. Comparative Values of Methods for measuring Velocities. Distribution of Velocities in a Channel. Mean Velocity deduced from Surface Velocity. Formulæ of Discharge:— Humphreys and Abbot's Formula; Darcy and Bazin's Formula; Ganguillet and Kutter's Formula. Concluding Remarks.

THE measurement of the discharge of a river is often necessary previously to the execution of works of improvement, and is also essential to a complete knowledge of the physical characteristics of a river. The flood discharge is required in designing works for the mitigation of floods, and the discharge at various periods is important for arranging irrigation works; whilst the low-water discharge is also needed for ascertaining the navigable capabilities of a river.

Measurement of discharge, moreover, forms the fundamental step towards the establishment of any formula of discharge, for calculating beforehand the discharge through a deepened river or new cut, and along an irrigation or drainage channel, so as to ascertain the effects of such works before they are undertaken, and in order to regulate the size and slope of the channel so that it may pass the desired volume of water.

Definition of Terms. The *discharge* of a river is the volume of water which it passes into, or towards the sea in a given unit of time. It is usually reckoned in cubic feet per minute or second.

The *slope* or *fall* of a river is the inclination of its water surface ; and it may be conveniently expressed either in feet and inches per mile, or in terms of the actual ratio of the fall to the length.

The *sectional area* of a river is the area of the cross-section of its channel taken at right angles to the current, and is expressed in square feet. When the actual, or possible discharge of a river is to be measured, only that portion of the cross-section is taken into account which is below the actual, or assumed water-line.

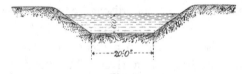

Fig. 5

The *wetted perimeter* is the portion of the boundary line of the cross-section of the channel below the water-line.

The *hydraulic mean depth*, or, as it is often termed, the *hydraulic radius*, is obtained by dividing the sectional area of the stream by the wetted perimeter of the channel. For instance, if a channel (Fig. 5) has a base of 20 feet, with slopes on each side of $1\frac{1}{2}$ to 1, a depth of water of 6 feet, and consequently a width at the surface of 38 feet, the sectional area of the stream is 29 feet $\times$ 6 feet $= 174$ square feet, and its hydraulic mean depth, or R,

$$= \frac{\text{sectional area}}{\text{wetted perimeter}} = \frac{174}{41 \cdot 6} = 4 \cdot 18 \text{ feet.}$$

In fact, R is the height of the rectangle whose base is the length of the wetted perimeter, and whose area equals the sectional area of the stream.

The *mean velocity* is the average velocity of all the elements of the current.

The discharge is obtained by multiplying the sectional area of the stream by the mean velocity.

Measurement of Cross-Sections. To ascertain the discharge of a river, it is necessary, in the first place, to find its sectional area. In gauging a river, as straight and uniform a reach as possible should be chosen, and where the flow appears to be regular, as this indicates that the section is fairly uniform. To ensure accuracy, several cross-sections should be taken along the part of the river selected. The cross-sections are obtained by taking soundings, at known intervals, in a straight line across the stream.

When the river is not very broad, a strong thin rope, previously wetted and pulled out, tightly stretched at right angles to the stream, from bank to bank, in the line of the proposed cross-section, serves conveniently as the base-line for the soundings. Where great accuracy is required, a wire should be employed in place of a rope. The positions of the soundings are measured along the rope, starting from the edge of the water on the right bank[1]. When the depth does not exceed 15 feet, the soundings are most conveniently taken by means of a round, slightly tapering, wooden pole, on which the lengths are marked from the bottom upwards. The pole must be terminated at the bottom by a broad flat piece, so that it may not sink into mud; and if the stream is strong, a little lead may be inserted at the bottom of the pole with advantage. The soundings are taken from a boat kept with its head up stream and just under the rope, so that the man taking the soundings may be able

---

[1] The right and left banks of a river are considered to be to the right and left, respectively, of a person when looking down stream.

to lower the pole in the line of the rope. When the river is too deep, or the current too strong, for using a pole for the soundings, a measuring chain with a weight at the end is employed.

It is advisable, if possible, to select a time for taking the cross-sections when the river is neither in flood nor very low. In flood-time there is frequently considerable difficulty in fixing the cord, in keeping the boat in position, and in taking the soundings. When the river is very low, there is less width of channel available for sounding, which is the easiest and quickest way of taking the measurements in the bed of the river, and there is liable to be too little water for floating the boat over the shallows.

The level of the water at the various places chosen for the cross-sections, and any alterations in level whilst the soundings are being made, are obtained, either by means of a spirit-level, or by reading off the height of the water on marked gauge-boards set in the water to a known datum. The measurements for the cross-sections are extended from the edge of the water over the banks on each side, for a certain distance, by levelling, so that the cross-sections may be known for varying heights of the river.

When a river is too broad for stretching a rope or wire across it, the line of the section is marked by erecting a pole on each bank; and a man is stationed at one of the poles. Directly the boat conveying the sounding party, slowly dropping down stream, crosses the line, a signal is made by the man on the bank, and the sounding is immediately taken, everything having been got in readiness beforehand with the sounding chain lowered nearly to the bottom. At the same instant, the exact distance of the sounding from the banks is noted, either by two observers, stationed with theodolites at the extremities of a measured base-line on one of the banks, who take the angles between the boat and the base-line, or by a person with a sextant in the boat taking the angles subtended by known objects on shore. Any required number of soundings can be thus taken along a particular line across the river, at different distances from the banks, from which the cross-section can be plotted.

Velocity Observations. Having obtained a sufficient number of cross-sections, and deduced from them the mean sectional area of the river or channel under examination, it is necessary to ascertain the mean velocity of the current in order to estimate the discharge of the river. There are three general methods by which this velocity may be obtained:

1. *Floats.*   2. *Current-Meters.*   3. *Gauge-Tubes.*

1. *Floats.* The simple expedient of throwing a floating substance into a stream, and observing the time it takes to traverse a measured distance, for ascertaining the velocity of the current, presents many difficulties in practical application. A floating body merely indicates the velocity of the portion of the current in which it happens to be situated; whereas the velocity of a stream varies, not only at different distances from the banks, but also at different depths in the same vertical line.

For surface velocities, an orange forms an excellent float, as its specific gravity is very little less than that of water, and it therefore floats almost wholly immersed; but pieces of wood, cork, or other easily procurable floating substances are often employed. Surface-floats, however, are useless in windy weather, as wind has a great effect on the top layers of water, so as even sometimes to make the surface water move in an opposite direction to a feeble current below, as observed by me during some float experiments in the Firth of Clyde[1].

---

[1] Minutes of Proceedings Institution C.E., vol. lxxi, p. 50.

The double float is employed for measuring sub-surface velocities at various points in a stream, so as to obtain the mean velocity of the current.   It consists of a small surface-float connected by a cord to a large float carefully weighted, so as to sink in the water and keep the connecting cord stretched, but not enough to pull the surface-float under water (Fig. 6).   The length of the connecting cord is varied according to the depth at which the velocity is to be measured : and whilst the current imparts its motion to the lower float, this motion is rendered apparent by the corresponding movement of the upper float.

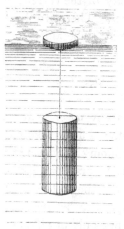

Double Float.
Fig. 6.

The extensive series of experiments on the flow of the Mississippi, by Messrs. Humphreys and Abbot[1], were made with double floats.   The lower floats were kegs, from 9 to 12 inches high, and 6 to 8 inches in diameter, without top or bottom, and ballasted with lead so as to sink and assume an upright position.   They were retained at the desired depth by surface-floats, to which they were attached by cords of suitable length.

Double floats were also used in measuring the flow of the Connecticut River[2], as well as two types of current-meters, by General T. G. Ellis.

The double float employed by Mr. Robert Gordon in gauging the flow of the Irrawaddi, consisted of two wooden cylinders, 6 inches in diameter : the lower float, 1 foot long, was weighted with lead, so that the surface-float,

[1] 'Report on the Physics and Hydraulics of the Mississippi River.' Captain A. A. Humphreys and Lieutenant H. L. Abbot, 1861.

[2] 'Report of the Surveys and Examinations of the Connecticut River.' General T. G. Ellis. 'Report of the Chief of Engineers, U. S. A.,' 1875, Part 2, p. 305, and Plate 1.

1 inch thick, to which the lower float was attached by a thin
cord, floated about three-fourths immersed [1].

In the Roorkee hydraulic experiments, conducted by Lieut.-
Col. Allan Cunningham on the Ganges Canal [2], two patterns
of double floats were adopted for measuring sub-surface
velocities. A sphere of wood, 3 inches in diameter and
weighted with lead, was connected by a thin brass wire to
a pine disk, 3 inches square and $\frac{1}{4}$ inch thick; and a copper
spherical shell, $1\frac{3}{8}$ inches in diameter, weighted with lead,
was connected by a very fine silk thread to a cork disc,
1 inch square and $\frac{1}{8}$ inch thick.

Tin tubes, 2 inches in diameter, loaded with lead so as to
float vertically, nearly reaching the bottom, were employed
by Mr. J. B. Francis in the measuring flumes at Lowell,
Massachusetts, and other places, for determining the amount
of water drawn from the canals [3]. The flumes were constructed in tolerably straight and uniform parts of the canals,
and rectangular in section, by lining the bottom and sides
with smooth planking, so that the current through them
might be as regular as possible; and the tin tubes indicated
the mean velocity of the vertical section of the stream in
which they floated. By means of these tube-floats, placed
at different distances from the banks, the mean velocity
across the whole cross-section of the stream can be
ascertained with considerable accuracy.

The velocity-rods used for determining the mean rate of
flow on the Ganges Canal were very similar: they consisted
of loaded wooden poles, 1 inch in diameter, and also 1-inch
tin tubes weighted at the bottom with a short piece of iron
rod, and closed at each end, the latter of the two proving
far the best [4].

---

[1] 'On the Theory of the Flow of Water in Open Channels.' Robert Gordon.

[2] 'Roorkee Hydraulic Experiments.' Captain Allan Cunningham.

[3] 'Lowell Hydraulic Experiments.' J. B. Francis.

[4] 'Recent Hydraulic Experiments.' Major Allan Cunningham. Minutes of
Proceedings Institution C. E., 1883, vol. lxxi, p. 20.

2. *Current-Meters.*   Several current-meters have been
designed for measuring the velocity of a current by the
number of revolutions performed, in a given time, by a screw
or vane, fixed on a horizontal axis in the front part of
the instrument (Figs. 7 and 8).   Woltmann's turbine current-
meter, with improvements by Mr. Baumgarten, is the form
of instrument generally used in France[1]; and it was also
used in gauging the Connecticut River, as well as a novel
type of meter with four conical-ended vanes, designed to
move with little friction and to avoid catching weeds[2].
The revolutions are recorded by toothed wheels, connected
with the axis of the screw, which turn an indicator revolving

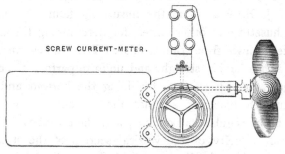

SCREW CURRENT-METER.

Fig. 7.

round a graduated dial.   The distance between each mark
on the dial is arranged to correspond with the space travelled
by the indicator when the current has run a certain distance,
usually taken at one foot.   One or more wings at the tail
of the meter keep it, when immersed, with the axis of the
screw in a line with the current.   A brake is fitted to the
instrument, so that the screw can be set free to revolve,
or can be stopped, by merely pulling tight or slackening
a string, above water, connected with the brake.

[1] Annales des Ponts et Chaussées, 1847 (2), p. 326; 1858 (1), p. 121; 1860
(1), p. 215; and 1883 (1), p. 219.
[2] 'Report of the Chief of Engineers, U.S.A.,' 1875, Part 2, pp. 306–309,
and Plates 2 and 3.

To measure the velocity of a current, after noting the reading of the index, the current-meter is immersed at the desired position in the river, and lowered to the required depth from a boat, being held in position by one or more strings or rods. The brake is then raised, and the time noted. At the close of the period of observation the brake is applied, the meter is lifted out of the water, and the distance traversed by the index is ascertained. If the index registers in feet, the number of divisions traversed by the

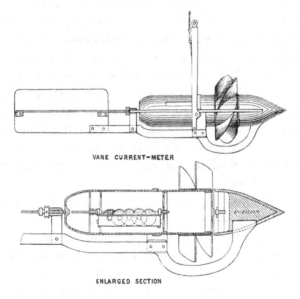

VANE CURRENT-METER

ENLARGED SECTION

Fig. 8.

index in the period, divided by the time, gives the velocity of the current at the point where the meter was held. If the mean velocity in a definite vertical plane is required, it is only necessary to lower the current-meter gradually, at a given spot, to the bottom, and then raise it again at a uniform rate.

Though the manufacturers profess to make the index register the exact distance run by the current, this is only approximately realized; and when very exact results are required the current-meter should be tested, either by placing it in a stream with a uniform known rate of flow, or by drawing it along at a given velocity in still water, in order to ascertain what each division of the index, or a definite number of revolutions of the screw, actually represent, and what allowance should be made for frictional resistance at different velocities.

In order to reduce the complication of toothed wheels which check the rotation, to detect any temporary stoppage of the rotation, and to avoid the necessity of lifting the current-meter out of water after each observation, an electrical contrivance has been fitted to the meter, which rings a bell above water when a certain number of revolutions have taken place. The same object has also been attained by making the axis of the screw move a small hammer in a definite number of revolutions. The hammer strikes a tightly stretched wire which passes up out of the water to a resonance box, so that each stroke of the hammer is distinctly audible [1].

3. *Gauge-Tubes.* In 1732, Pitot proposed to gauge the flow of water with a glass tube, having a short bend at one extremity at right angles to the tube, and open at both ends. The tube being immersed vertically in a current, with the bend horizontal and its end facing the current, the water in the tube is raised by the pressure of the current; and the height of the column of water above the surface is proportionate to the velocity of the current, which is represented by the formula,

$$V = \mu \sqrt{2gh},$$

where h is the height of the column, and μ is a coefficient which can be determined by drawing the tube through still water at known rates. In this simple apparatus, the height of the column was difficult to determine, owing to the ripples

---
[1] 'Deutsche Bauzeitung,' 1880, p. 229.

on the surface of the stream and the oscillations inside the tube, and was, moreover, not quite an exact measure of the velocity.

To obviate these defects, Mr. Darcy employed two tubes (Fig. 9), ⅜ inch in diameter, connected at the top by a copper tube in which a stopcock is fastened, which enables communication to be made at pleasure with the outer air [1]. Copper tubes are fastened at the bottom of each of the glass tubes, and bent at right angles to the glass tubes, and to each other, for a length of 6 inches, and terminated by mouthpieces having orifices of only $\frac{1}{17}$ inch diameter. The lower copper tubes have a double-acting stopcock which opens or closes them simultaneously. The instrument is lowered into the water with both stopcocks open, unless it has to be put right under water, in which case the upper stopcock is closed before the instrument is submerged; and the tubes are immersed to the desired depth, with one mouthpiece pointing to the current and the other at right angles

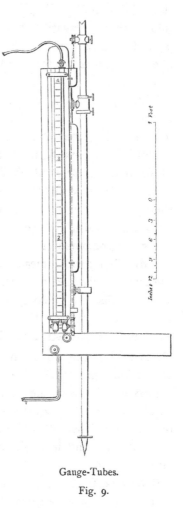

Gauge-Tubes.

Fig. 9.

to it. The water flowing into the tubes rises highest in the

[1] Annales des Ponts et Chaussées, 1858 (I), p. 354; and 'Recherches Hydrauliques,' Darcy and Bazin.

tube pointing to the current; the lower stopcock is then closed, the instrument is raised, and the difference in level of the water in the two tubes is measured at leisure.

It is important to see that the mouthpieces are clear, that the tubes are kept perfectly vertical, and to wait for taking the observation till the oscillations of the column of water have ceased. It is also advisable to repeat the operation three or four times, and take the mean of the results.

This apparatus was used by Messrs. Darcy and Bazin in their experiments on the canals of Chazilly and Grosbois, for measuring the discharge through various channels, with the object of investigating the laws of the flow of water in open channels, and obtaining a general formula of discharge. Improvements have been recently introduced in the apparatus by Mr. C. Ritter[1], and the instrument has been frequently used in France.

Hydrodynamometer. An ingenious contrivance, designed by Mr. de Perrodil, measures the velocity of a current by the torsion produced on a wire by the pressure of the water against a disc[2]. The disc is fixed at the end of a horizontal arm projecting from the extremity of a brass wire placed vertically in the stream. The velocity of the current is obtained by the equation $V = c \sqrt{a}$, where a is the angle of torsion; and the value of c can be determined, either approximately from the coefficient of torsion of the metal employed, or actually by experiment. The instrument used by Mr. de Perrodil indicated a velocity as low as $\frac{3}{8}$ inch per second; and it could be constructed so as to be still more sensitive.

Comparison of Methods for measuring Velocities. Experimenters have expressed considerable differences of opinion as to the correctness of the methods employed for

---

[1] Annales des Ponts et Chaussées, 1885 (1), p. 1058.
[2] Ibid. 1877 (1), p. 467, and 1880 (1), p. 11.

measuring the velocity of a current. Messrs. Humphreys and Abbot discarded the current-meter, as unsuitable for the deep, turbulent, and irregular current of the Mississippi, and relied exclusively on double floats; and on the Connecticut River, double floats were found the most reliable method for measuring sub-surface velocities in a uniform channel. Lieut.-Col. Allan Cunningham tried three patterns of current-meters on the Ganges Canal, but abandoned them in favour of velocity-rods, on account of the insurmountable difficulties he found in their use. On the other hand, double floats were considered worthless by Mr. Révy, who trusted entirely to current-meters, of the type shown in Fig. 7, for gauging the discharge of the Paranà and La Plata rivers[1]; whilst Mr. Darcy was adverse to double floats, and resorted to gauge-tubes; and Professor Unwin has expressed his opinion in favour of current-meters, as more correct and serviceable than floats[2]. Mr. Bazin also, after comparing the results of numerous experiments with double floats and current-meters in large streams, came to the conclusion that, in many cases, the use of double floats led to serious errors[3].

The main objections urged against double floats are, that the motion of the lower float is not accurately indicated by that of the float at the surface, and that, moreover, the upper float and connecting cord, being acted upon by a different portion of the current, modify the motion of the lower float. Also, that it is impossible to ascertain whether the lower float remains at the desired depth, and is not lifted by eddies into a higher part of the current, and constantly shifting its position. Mr. Gordon observed, on the large river Irrawaddi, that the motion of the top float was too uniform to allow of the existence of the erratic movements of the

[1] 'Hydraulics of Great Rivers: The Paranà, the Uruguay, and the La Plata Estuary.' J. J. Révy.

[2] Minutes of Proceedings Institution C. E., vol. lxxi, p. 40.

[3] Annàles des Ponts et Chaussées, 1884 (1), p. 554.

lower float attributed to it by the opponents of the method; and my observations of the motion of the lower float in small clear streams led me to a similar conclusion.

The tube-float is free from these objections; and no divergence from the perpendicular can occur without being manifested by the inclination of the portion of tube above the water. The tube-float also measures the mean velocity of the current in the plane in which it floats, and thus is more rapid, as well as more accurate in its measurements than the double float. The tube-float, however, is not suitable for very deep rivers, or for channels where the depth varies considerably, or where weeds grow in the bed; and all floats are liable to deviate from a straight course. Mr. Francis has stated that the tube-floats in the Lowell experiments travelled a little faster than the current; whereas Major Cunningham has calculated that such floats move somewhat slower than the water in which they are immersed. Except for the opposition offered to the air by the portion of the tube out of water, there seems to be no reason why the average velocity of the float and the current should not be identical; though with a change of velocity in the current, the momentum of the float would prevent its assuming at once the same change in rate.

Besides their applicability to large turbulent rivers like the Mississippi, floats are valuable for measuring the flow of sluggish streams in which the ordinary current-meter would not act; and, with the exception of the hydrodynamometer, which has hitherto been only tried experimentally, floats furnish the only available method in such cases.

Current-meters are merely inferential recorders of velocity, and depend for their correctness on the accuracy of their mechanism. Silt and floating weeds are liable to derange them; and the increased influence of friction at low speeds renders them unsuitable for measuring feeble currents. Their registration, however, can be checked by using two instruments, and the continuity of the revolutions proved by an electrical recorder[1]. Current-meters possess the advantages, of not necessitating a long, uniform, unimpeded channel for the observations, of restricting the measurements to a single well-defined cross-section of the river, of being easily placed and kept at the precise positions in the current at which the velocities are to be noted, and of enabling the observations to be taken much more rapidly than with floats. Moreover, they can measure velocities closer to the bottom than floats.

Darcy's gauge-tube gives good results in small channels, with depths not exceeding 6 feet, and when the stream has a fair velocity; but it would be inapplicable to large rivers, and very low rates of flow.

The sensitiveness of the hydrodynamometer, rendering it suitable for measuring very feeble currents, appears to furnish its chief claim to more extended adoption. Current-meters and tube-floats are, in general, the best instruments for measuring the velocity of flow; whilst double floats and gauge-tubes may occasionally be employed with advantage. The choice of method, in any particular case, must depend upon the size and configuration of the channel, the nature of the flow; and the general condition of the stream, and should be made in accordance with the indications given above. The checking of the results by the occasional use, where practicable, of a second method of observation, is always valuable, both in securing a greater degree of accuracy and more certainty as to the correctness of the experiments, and also in extending experience as to the relative scope and reliability of each method.

As the discharge of a river undergoes frequent changes, it is generally sufficient, in estimating its flow at different periods or heights, to obtain by experiment an approximate

[1] 'Die Messungen in der Elbe und Donau, und die hydrometrischen Apparate und Methoden des Verfassers.' Professor von Harlacher.

measurement of its velocity; but when the amount of water supplied for any purpose has to be calculated, or still more when the observations of velocity are to be used as a basis for a general formula of discharge, the utmost care and exactness are required in measuring the flow.

Distribution of Velocities in a Channel. The bottom and sides of a channel necessarily retard the flow of the water close to them, in proportion to their roughness; and this retardation, in irregular channels, is more due to the impeding of the flow by eddies than by friction alone. The air also in contact with the surface of the stream has a similar, though naturally slighter influence, unless the wind happens to be blowing in an opposite direction to the current. Accordingly, the maximum velocity of flow in a river, along a straight reach, is found in the central portion of the stream, and generally somewhat below the surface, the actual position depending on the size and condition of the channel and the velocity of flow. The velocity increases at first from the surface downwards for a short distance, and then decreases down to the bottom where it reaches its minimum. Messrs. Humphreys and Abbot found that the variation in velocity of the flow of the Mississippi, in the place where their experiments were conducted, at different depths on the same vertical, might be represented by a parabola, having its axis parallel to the surface, and at the depth below the surface of the position of maximum velocity, the abscissæ representing the velocities at the different depths, and the ordinates the vertical distances of these depths from the line of maximum velocity. In the Mississippi experiments, the maximum velocity was, on the average, nearly one-third of the whole depth below the surface, varying with the direction of the wind. Mr. Bazin had previously shown that the velocities of the current, in small channels, could be represented by the abscissæ of a parabola with a horizontal axis; and he subsequently indicated the approximation of the velocities on various rivers, at different depths, to the parabolic form[1], the curve varying with different channels, and with the position of maximum velocity. Professor von Wagner, in investigating the vertical curves of velocities of large and small rivers, also found that they corresponded approximately with parabolas having their horizontal axis at the depth of the maximum velocity, but that they deviated from the parabola towards the bottom, and near the surface, and that the depth of the maximum velocity varied from a little below the surface to a little over a fourth of the full depth[2].

The horizontal curves of velocities do not exhibit an approximation to any definite curve; for the velocities vary very little in the main stream, diminishing very little from the centre towards the sides, till on approaching the banks they are rapidly reduced.

The retardation of the surface velocity has been attributed, by some investigators, to the rising of the lower water to the surface, being checked in its flow by striking against the rough bottom and sides of the channel. Water, however, rising from the bottom would have to traverse the layers of maximum velocity before reaching the surface, and would check the lower layers rather than the surface layer. The checking of the flow against the sides might, indeed, tend to raise the water-level near the banks, resulting in a transverse flow towards the centre, which would retard the surface layer, and to which Mr. F. P. Stearns has attributed the reduction of the surface velocity[3]. The existence, however, of this transverse surface flow in rivers would have to be clearly established before it could be considered as aiding the action of the air and wind in checking the surface flow.

Mean Velocity deduced from Surface Velocity. Gauging

[1] Annales des Ponts et Chaussées, 1875 (2), p. 309, and Plate 26.

[2] 'Hydrologische Untersuchungen an der Weser, Elbe, dem Rhein, und kleineren Flüssen.' Professor von Wagner.

[3] Transactions of the American Society of Civil Engineers, vol. xii, 1883, p. 331.

operations would be greatly simplified and shortened if the mean velocity could be arrived at by merely measuring the maximum surface velocity.

With the object of determining the ratio between the maximum surface velocity and the mean velocity, De Prony made some experiments in wooden troughs, and Messrs. Baldwin, Whistler, and Storrow in channels lined with planks. The coefficient obtained for converting the observed surface velocity into mean velocity, were ·816 by De Prony, and from ·810 to ·847 by the latter observers in different channels. Subsequent experiments indicate that the co-efficient is generally comprised within the limits of o·8 and o·9, depending upon the size of the channel and the nature of its bed. In the Mississippi experiments, the coefficient exceeded o·9 ; but it is probable that the influence of the long connecting cord and surface-float caused too large values to be recorded for the velocities towards the bottom, and thus gave too high a value to the mean velocity; for Mr. Robert Gordon, in checking his experiments on the Irrawaddi with a current-meter, obtained considerable reductions in the velocities on approaching the bottom, compared with those recorded by double floats [1]. The coefficient would be greatest for large deep rivers with smooth, uniform channels, and least for small shallow streams with rough beds.

Messrs. Darcy and Bazin derived from the results of their experiments the following formula, giving the relation between the maximum velocity and mean velocity :

$$U - V = 25 \cdot 36 \sqrt{RS} \ ;$$

where U is the maximum velocity in feet per second, V the mean velocity, R the hydraulic radius in feet, and S the slope. Assuming an average value for V of $90\sqrt{RS}$, we

[1] 'Notes on Subjects connected with Works in the Irrawaddy Circle, British Burma, with Records of Experiments on the Double Float and Woltmann Meter Current Measurements,' 1883.   Robert Gordon.

obtain $U = 115\cdot36\sqrt{RS}$, and therefore $U \times \cdot78 = V$, correctly giving a somewhat smaller coefficient for converting the maximum velocity into the mean velocity than that given above for the maximum surface velocity.

Formulæ of Discharge. One of the great aims of all hydraulicians who have investigated the flow of water in channels, has been to obtain a general formula, applicable to any kind of channel and rate of flow, from which the discharge can be readily calculated, provided the section of the channel and the slope are known. Such a formula is specially required when the channel of a river has to be enlarged, or a straight cut made, and in designing drainage and irrigation canals, so that they may discharge the desired volume of water within the given time, and at a rate not liable to be injurious to the bed and banks of the channel. In existing watercourses, the direct measurement of the discharge is much to be preferred to the results of any formula.

Formulæ of discharge are always given in terms of the mean velocity, for when once the velocity for a given section and slope has been ascertained, it is only necessary to multiply the velocity by the sectional area of the channel, to obtain the discharge, $D = VA$. Owing to the variable flow of water in ordinary channels, a velocity formula cannot be arrived at by any strictly mathematical process; and therefore all the formulæ obtained are empirical, being derived from the results of experiments, and their only claim for acceptance being based on their accordance with numerous experimental results arrived at with the utmost precision. As the reliability of a formula of discharge depends on the accuracy of the experiments on which it has been founded, only those experiments should be taken into consideration, in preparing a general formula, the correctness of which has been fully established. With this reservation, a formula of discharge will be suited for general adoption in proportion as it agrees with the measured discharges of very different channels under different conditions of flow.

It is evident that the velocity of a current increases with any increase of its slope, or of its hydraulic radius; and therefore S and R must enter into any formula expressing the value of V, the mean velocity. Coulomb had concluded from his investigations that the resistance of the wetted perimeter to the flow comprised two functions, one varying with the velocity, and the other with the square of the velocity, from which Prony derived the formula,

$$a\,V + b\,V^2 = RS,$$

in which a and b were friction coefficients to be obtained experimentally. Subsequently Chézy proposed to omit the first term as insignificant for a very small channel with a fair velocity of flow, and established the simple formula,

$$V = c\sqrt{RS},$$

which was accepted by many hydraulicians, who merely introduced modifications in the value of the assumed constant c. Downing adapted it to British measures by making $c = 100$; so that it took the form $V = 100\sqrt{RS}$, where V is the mean velocity in feet per second, R the hydraulic radius in feet, and S the slope. Some of those who adopted this formula gave different values to c according to the size of the stream, the smallest value assigned to it being 68, for small streams; but none of them assigned it a larger value than Downing's[1], which is applicable to a discharge of about 17,000 cubic feet per second. Most of the more recently proposed formulæ retain the general form of Chézy's formula, but introduce modifications in the coefficient c, which has been made variable; whereas in the earlier formulæ, no modification of the co-

---

[1] 'Hydraulic Manual.' L. D'A. Jackson, p. 252.

efficient was introduced for variations in slope and hydraulic radius, or for the degree of roughness of bed.

Humphreys and Abbot's Formula. Though the object of Messrs. Humphreys and Abbot's investigations was to ascertain the flow of the Mississippi by direct measurement, with a view to the regulation of the river, they saw the importance of endeavouring to deduce a general formula of discharge from the results of their observations on one of the largest rivers of the world, to serve as a guide in determining the flow in very large channels. Having found that the maximum velocity was below the surface, they deemed it necessary to add the surface line to the wetted perimeter, as equally retarding the flow; so that they replaced the hydraulic radius, $R = \dfrac{A}{P}$, which appears in all the other formulæ, by

$R' = \dfrac{A}{P + W}$, or the area of he cross-section divided by the entire perimeter of the section, of which W is the width at the surface of the water, and P the wetted perimeter. They also introduced a coefficient into their formula, containing the hydraulic radius R, and therefore changing with variations in R, namely $m = \dfrac{1 \cdot 69}{\sqrt{R + 1 \cdot 5}}$; and the formula in its final shape became,

$$V = \left\{ \left(0 \cdot 0081 \ m + \sqrt{225 \ R' \sqrt{S}} \right)^{\frac{1}{2}} - 0 \cdot 09 \sqrt{m} \right\}^{2}.$$

In this complicated form, it is difficult to compare this formula with others; but omitting the small first and third terms, and assuming W equal to P, and substituting a function f, varying with R, in the remaining middle term in place of the omitted terms, we obtain a simplified equation, $V = 10 \cdot 6 \ f \ \sqrt{RS^{\frac{1}{2}}}$, in which it is evident that the main difference between this and the earlier formula, $V = c \sqrt{RS}$, is that the velocity varies with the fourth root of S instead of the square root, and to some extent with changes in R. The modification introduced with regard to S makes the new formula only applicable to rivers with very slight slopes; and the absence of any variation in the coefficient according to the nature of the bed, though of comparatively little importance in a large river like the Mississippi, renders it unsuitable for smaller streams, where the nature of the bed in a restricted cross-section greatly affects the flow. Moreover, the important alteration in the value of the hydraulic radius, by the introduction of W, does not appear to be warranted, either by the general moderate depth below the surface of the position of maximum velocity, or by the friction of the surface water against the air compared with the friction occasioned by the bed; and if the reduction in the surface velocity is attributable to the roughness of the sides, the coefficient should vary according to the amount of this roughness.

It is, accordingly, evident that the Mississippi formula, whilst valuable as throwing light on the conditions of flow in large rivers with gentle slopes, is unsuitable for general adoption. Nevertheless, the Mississippi experiments have the special value of furnishing a test of the applicability of other general formulæ of discharge to the case of rivers like the Mississippi.

Darcy and Bazin's Formula. The experiments inaugurated by Mr. Darcy, and continued after his death by Mr. Bazin, were undertaken with the object of ascertaining the influence which the condition of the lining of a channel has upon the discharge. The results of these experiments in small, regular channels were embodied in a formula,

$$V = \sqrt{\dfrac{1}{a + \dfrac{\beta}{R}}} \ \sqrt{RS,}$$

in which a and β are given different values according to the degree of roughness of the bottom and sides of the channel[1]. Thus a varies from ·000046 for cement or planed wood, up to ·00085 for earth; and β varies from ·0000045 up to ·00035 for the same differences of lining.

This formula, accordingly, whilst resembling Chézy's formula in its general shape, introduces variations for the roughness of the bed of the channel, as well as for changes in R allowed for in the Mississippi formula, the constant c being replaced by the variable coefficient $\sqrt{\dfrac{1}{a+\dfrac{\beta}{R}}}$. The formula naturally presents a considerable contrast to the Mississippi formula, as the two sets of experiments on which these formulæ were based were conducted under diametrically opposite conditions, representing the two extremes of such measurements. In the Mississippi, changes of slope had great influence on the discharge; whilst in Mr. Bazin's observations, the roughness of the channel exercised a very notable effect.

Comparatively recently Mr. Bazin has revised his original formula by the aid of further experiments in small earthen channels and canals, and on the Seine and the Saône[2], arriving at the expression,

$$V= \left\{ \frac{1}{·00008534\left(1+\dfrac{4·1}{R}\right)} \right\}^{\frac{1}{2}} \sqrt{RS}.$$

As the roughness of the bed, composed of earth, is supposed to be similar in these cases of ordinary rivers and streams of moderate dimensions and having a steady flow, the coefficient simply varies with R; and when $R=5$ feet, $V=80·2\sqrt{RS}$; when $R=10$ feet, $V=91·1\sqrt{RS}$; and when $R=30$ feet, $V=101·5\sqrt{RS}$.

[1] 'Recherches Hydrauliques,' H. Darcy and H. Bazin, p. 125; and Annales des Ponts et Chaussées, 1871 (I), p. 22.
[2] Annales des Ponts et Chaussées, 1884 (I), p. 587.

For torrential rivers carrying along shingle, the value of the coefficient has to be modified, so that the formula becomes,

$$V = \left\{ \frac{1}{\cdot 000122 \left(1 + \frac{5 \cdot 74}{R}\right)} \right\}^{\frac{1}{2}} \sqrt{RS};$$

and under these conditions, when $R = 5$ feet, $V = 61 \cdot 8 \sqrt{RS}$; when $R = 10$ feet, $V = 72 \cdot 2 \sqrt{RS}$; and when $R = 30$ feet, $V = 82 \cdot 9 \sqrt{RS}$. Observations have shown that the value of the coefficient is 65 2 for the torrential tributaries of the Upper Loire, and 68·8 for the Rhine at Basle, in spite of the fair size of the river, the bed being covered with large shingle.

The values of the coefficient given above for certain values of R are very nearly comprised within the range of values assigned to c, in the formula $V = c \sqrt{RS}$, by the earlier hydraulicians. Instead, however, of the coefficient being considered constant, and having a special value assigned to it by each experimenter according to the streams observed, the above coefficient varies with R for channels having ordinary, regular, earthen beds, and is changed when the nature of the bed is different.

Ganguillet and Kutter's Formula.   The formula of Messrs. Darcy and Bazin, though very valuable for estimating the discharge of small channels, and based upon correct scientific principles, was not applicable to large irregular rivers with very slight slopes, like the Mississippi. Moreover, Messrs. Humphreys and Abbot expressed a desire that the suitableness of their formula for rivers with considerable slopes should be tested by experiments. These circumstances led Messrs. Ganguillet and Kutter to gauge the flow along mountain channels built of rubble masonry, with a semicircular cross-section and a considerable slope, serving to discharge flood waters; and finding that the Mississippi formula was inapplicable to these conditions, they proceeded to construct a formula capable of embracing the results of the Mississippi gaugings, as well as their own, Mr. Bazin's, and other observations.

Starting from Messrs. Darcy and Bazin's formula as a basis, they eventually found it expedient to modify the form of the coefficient, so that the formula became, $V = \dfrac{y}{1 + \dfrac{x}{\sqrt{R}}} \sqrt{RS}$,

where y and x vary with the slope[1]. This introduction of the slope into the coefficient was rendered necessary by the anomalous fact that, whilst c decreased with a decrease of the slope in small channels where the bed was not very rough, it increased with a decrease of the slope in the Mississippi results. This change was found to occur when R passed the value of about one metre, or 3·281 feet, though the actual value of R at which the change took place varied with the roughness of the bed. This anomaly has been attributed to the intensification of the conflicting currents and eddies by an increased slope in large rivers with very irregular beds, and in small streams with very rough channels.

In pipes and small channels with steep slopes, where the influence of a variation of the slope on the coefficient may be neglected, making $y = \dfrac{l}{n} + a$ and $x = an$, where $l = \sqrt{R'}$ for the special value $R' = 3\cdot281$, and therefore $l = 1\cdot811$, $a = 41\cdot6$, and n the coefficient of roughness varies between ·009 and ·04, the equation becomes,

$$V = \frac{\dfrac{l}{n} + a}{1 + \dfrac{an}{\sqrt{R}}} \sqrt{RS} = \frac{\dfrac{1\cdot811}{n} + 41\cdot6}{1 + \dfrac{41\cdot6n}{\sqrt{R}}} \sqrt{RS}.$$

For the general formula, applicable to any sized channel and any slope met with in practice a factor containing the

---

[1] 'A General Formula for the Uniform Flow of Water in Rivers and other Channels.' E. Ganguillet and W. R. Kutter. Translated from the German by R. Hering and J. C. Trautwine.

slope is introduced into the coefficient by substituting $a+\dfrac{m}{S}$ for a, where $m=\cdot 00281$ is the constant of a hyperbola used in constructing the formula.   The formula thus becomes,

$$V=\frac{\dfrac{1\cdot 811}{n}+41\cdot 6+\dfrac{\cdot 00281}{S}}{1+\left(41\cdot 6+\dfrac{\cdot 00281}{S}\right)\dfrac{n}{\sqrt{R}}}\sqrt{RS}$$

The authors of this complex formula claim that not only is it applicable to large rivers and small streams, but that it actually more nearly represents the results of the Mississippi gaugings than Messrs. Humphreys and Abbot's formula. They also assert its simplicity for practical use; but this is only attained by the aid of tables or graphic methods. Its complexity is a necessary result of its comprehensiveness; and in spite of this, it has met with very general approval as a skilful attempt to condense all varieties of flow into a single formula. The formula has come into use in Germany and Italy; and several persons who have investigated it have testified to its reliability. Lieut.-Col. Cunningham has stated that it agreed more nearly with his gaugings on the Ganges Canal than Messrs. Darcy and Bazin's formula. Mr. Bazin, however, has indicated that the only discrepancy between the two formulæ is found in cases like the Mississippi[1]; and he considers that the estimated velocities in the lower depths of the Mississippi require important corrections, owing to the influence of the current on the long connecting cord, as was proved in the later experiments on the Irrawaddi

The permanence of this formula in its present shape will depend upon the correctness of the gaugings of large rivers with slight slopes; for if these should prove to need important modifications, the formula, in so far as it is based upon them, will require alteration. The change of the influence of a variation of the slope on the coefficient beyond a certain

---

[1] Annales des Ponts et Chaussées, 1871 (1), p. 43.

value of R especially needs further investigation, for it presents an anomaly which is not clearly intelligible, and which more extended observations might either elucidate or show to be incorrect.

Concluding Remarks. The only way by which the time and trouble involved in gauging operations could be diminished would be by establishing a more exact relation, under definite conditions, between surface velocity and mean velocity, or between maximum˙velocity and mean velocity : for then a single set òf observations with a surface-float in the centre of the stream on a still day, or with a current-meter at the point of maximum velocity below the surface, would suffice for obtaining the discharge. Rapid and approximate methods of gauging the flow of rivers are very valuable in studying the physical characteristics of a river with a view to works of improvement, as it rarely is possible to resort to such elaborate investigations as those carried out with public funds on the Mississippi and the Ganges Canal.

The hydraulician is interested in observing the flow of water under the most diverse conditions, with the object of investigating the laws which govern its motion ; but the practical engineer has more limited aims. The use of a formula of discharge is absolutely essential for determining the discharge of a proposed river enlargement or new cut, or of a contemplated drainage or irrigation canal ; but such channels are made regular, and with fairly smooth sides and bottom, and bear no resemblance to the irregular channels of large rivers. Accordingly, great as is the interest attaching to the establishment of a general formula of discharge applicable to all known rivers and channels, a simpler formula of more limited scope would generally fulfil the requirements of engineers.

The most suitable formulæ of discharge for general use, namely those of Messrs. Darcy and Bazin, and Messrs. Ganguillet and Kutter, resemble in form the formula adopted by most of the earlier hydraulicians, $V = c\sqrt{RS}$, with the difference that c is variable instead of constant.   The variation of c directly with R, and inversely with the increase in roughness n, adopted in both these formulæ, is clearly logical, for the proportionate influence of the wetted perimeter in retarding the flow would become less in larger rivers, and a greater degree of roughness of the bed would diminish the velocity The effect of changes of slope on the value of c has not been decisively determined ; and whatever this effect may be on the very irregular channel of the Mississippi, it would not apply to a comparatively small regular, artificial channel. Accordingly, in ordinary cases, the amended formula of Mr. Bazin for channels in earth, or Messrs. Ganguillet and Kutter's formula in its simpler form, with $n = \cdot025$ to $\cdot03$[1], ought to suffice for estimating the probable discharge in designing an improved river channel, or a drainage or irrigation canal.

---

[1] The value of $n = \cdot025$ is for brooks or rivers flowing in channels of earth; whilst $n = \cdot03$ applies to channels containing detritus or aquatic plants.

CHAPTER III.

REGULATION AND CANALIZATION OF RIVERS.

Imperfect natural Condition of Rivers. Variable Flow of Rivers. Progression of Detritus. Highest and Lowest Water-levels to be observed. *Regulation of Rivers:*—Removal of Shoals and Obstructions; Contraction of Channel; Cross Jetties; Longitudinal Training Banks; Protecting and Easing Bends; Straight Cuts; Submerged Cross Dykes; Influence of Descent of Detritus; Remarks on Regulation of Rivers; Instances of Regulation Works. *Canalization of Rivers:*— Primitive Navigation; Stanches for Flashing; Locks, Weirs, and Level Reaches; Instances of Canalized Rivers; Remarks on Canalization of Rivers.

A RIVER to be perfect for navigation should be tolerably straight, without sharp bends, with a fairly uniform breadth and depth, and consequent regular flow, with neither a scarcity of water in dry weather, nor too full and rapid a flow in flood-time. Such a type of river is seldom found. Most rivers, in their natural state, present constantly varying conditions of breadth and depth, alterations in fall and flow, and a more or less winding course. When the fall increases abruptly, and the bed is rocky, rapids occur; and when the fall decreases considerably, and detritus consequently accumulates in the channel, the river becomes shallow, and widens out by eroding its banks in order to maintain its discharge.

Variable Flow of Rivers. Torrential rivers are not generally well adapted for navigation, on account of the great variations in their discharge, and consequently in their depth, at different periods of the year, and owing also to the rapidity of their flow, which presents a serious impediment to up-stream traffic.

E

The maximum discharge of the Loire at Briare has been estimated to amount to 300 times its discharge at its lowest stage; whilst the ratio of the two discharges is over 100 to 1 for the Dordogne at Libourne and the Moselle at Metz, and 90 to 1 for the Garonne at Langon. Though the Rhone is a torrential river, its flow is moderated by the Lake of Geneva, and by some of its tributaries not being of glacier origin, and therefore not having their greatest flow in the hot weather; so that in spite of its rapid fall, the ratio of its extreme discharges at Lyons is only 40 to 1, considerably less than the 70 to 1 of its tributary the Saône; whilst lower down at Valence, the ratio is only 30 to 1. The ratio of the extreme discharges of the Seine at Paris is 37 to 1, owing to the influence of its torrential tributaries the Yonne and the Marne; but it is reduced further down.

The variation in the discharges, which is excessive in torrential streams near their sources, becomes gradually less as the area drained increases; and the flow of rivers also generally becomes more gentle as they descend their valleys, on account of the reduction in the general slope of the land. As a reduction in fall necessitates a larger channel for conveying the same volume of water, and since the discharge increases, and becomes more regular as the various tributaries, from districts exposed to different meteorological conditions, join the main stream, a river is more suitable for navigation along the lower portion of its course. This, moreover, is indicated by the reduction in the ratio of the extreme discharges of large rivers at a distance from their sources; for instance, the ratio is 13 to 1 for the Rhine at Kehl, and less lower down; it is 8 to 1 for the Danube just above its delta, and only 4 to 1 for the Mississippi near the Gulf of Mexico.

The velocity of the current of a river not only varies for different heights of the river, but also exhibits frequent changes at different places along the river, owing to variations

in fall or alterations in depth. The variations in fall are due
to changes in the general slope of the land, or to hard or
rocky ridges barring the channel. The alterations in depth
result from differences in the erosive nature of the bed, from
modifications in the width of the channel, or from the serpentine course of the river (Plate 3, Fig. 1).

Progression of Detritus. The shifting nature of the bed
of most rivers must be taken into account in any works of
improvement. A river in its natural condition may present
much the same appearance from year to year; but the
shingle, gravel, sand, or silt forming the bottom of its bed is
always being carried down during high stages of the river,
and renewed by fresh supplies from above. Just as the
travel of shingle and sand along the sea-coast, under the
influence of the prevalent winds, is hardly perceived till
arrested by the erection of a projecting groyne, on the windward side of which the material accumulates; so the gradual
movement of detritus down a river is not readily noticed till
a modification of the channel disturbs the existing equilibrium.
The matter in suspension is, indeed, apparent; but the larger
and heavier materials, being rolled along the bed, escape
observation in the turbid current, and often form a considerable portion of the solid discharge. The noise, however,
of rolling shingle may be heard in a rapid torrent by attentive
listening; the movement of particles of sand may be perceived
in a clear, shallow, quickly-flowing current; and soundings in
the Mississippi at Helena and elsewhere have shown that
sand-waves gradually proceed down the river[1]. The vast
amount of material thus brought down in past times is
evidenced by the wide-spreading alluvial plains through
which many rivers flow, and by the extensive deltas found
near the outlets of several large rivers; whilst the continuance
of the same action at the present time, on a large scale, is

[1] 'Report of the Chief of Engineers, U. S. A., 1879,' part 3, p. 1963; and 1883,
part 3, p. 2216.

indicated by the sediment deposited by floods over considerable areas, of which the annual fertilizing deposit of mud by the Nile in Egypt is a notable instance, and by the steady progression of deltas, as proved by careful surveys of the outlets of the Rhone, the Danube, the Mississippi, and other delta-forming rivers. Though detritus is mainly brought down by floods, the scour is so much increased by the volume and velocity of the discharge, that, in spite of the accession of large supplies of fresh material, a river is generally deeper in flood-time than at a low stage.

Highest and Lowest Water-Levels. There are three levels of a river which it is specially important to note, namely, the highest known level, the highest level at which navigation can be carried on, and the lowest known level. A knowledge of the highest recorded flood level is necessary so that works which must not be submerged may be raised above that level, and that due protection may be afforded to low-lying lands adjacent to the river. All works connected with the navigation of a river have to be regulated in accordance with the highest navigable level. The lowest level indicates the minimum depth of water that is found in a river. The low-water level, however, upon which works of improvement are based, is not necessarily the lowest level to which the river may fall. The absolute lowest level of a river depends upon the combined effect of several physical causes which may not occur together, except at very rare intervals; and, accordingly, it is not generally advisable to provide, by more expensive works, for a very rare occurrence of short duration. The lowest ordinary water-level is usually selected as the datum below which the desired depth is to be maintained.

REGULATION OF RIVERS.

Removal of Shoals and Obstructions. To render a river suitable for navigation, its depth and flow should be made fairly uniform, by lowering the shoals to the requisite

minimum depth, and by regulating the rapids. A simple
expedient, at first sight, for accomplishing this object, would
be to dredge away the shoals, and to remove the obstructions causing rapids in the channel. Unless, however, the
shoals are hard permanent ones, which the river has been
unable to erode, they will be soon formed again by fresh
detritus, owing to the enlargement of the sectional area of the
channel by the dredging, and the consequent slackening of
the current. The removal also of the obstructions forming
rapids, whilst improving the flow at those parts, lowers the
water-level of the river above, by the increased facility of
discharge, and therefore diminishes the available depth when
the river is low, and causes the shallowest parts of the upper
channel to become fresh impediments to navigation, and,
moreover, increases the fall of the next rapid above. Accordingly, such methods of improvement are liable to prove
nugatory in the one case, and only serve to shift the position of
the impediments to navigation in the other, if supplementary
works are not carried out. Whilst, however, it is practically
useless to attempt to lower soft shoals permanently by
dredging alone, in rivers carrying along large quantities of
detritus, hard shoals, rocky reefs, and other obstructions
must be removed in order to effect improvements in a river;
and any undue lowering of the low-water level above must
be counteracted by further works.

Contraction of Channel. The only permanent method of
lowering soft shoals, whether caused by an accumulation
of detritus from a reduction in fall, or resulting from the
unstable direction of the main current between two bends,
is by reducing the width of the channel along the site of the
shoal, thus producing its removal by the increased scour,
which also prevents its forming again. This reinforcement
of the current where its scouring capacity is defective, secures
the maintenance of the increased depth by natural means;
but any contraction of the channel beyond the limits abso-

lutely necessary for securing the requisite depth, must be avoided, as the additional increase in depth would lead to a lowering of the low-water level to the detriment of the channel above, and would cause the deposit of sediment in the wider channel below.

Hard shoals and reefs have to be removed by artificial means; but their removal may render a contraction of the channel necessary, if the channel has been thereby unduly enlarged, so as to prevent silting or any serious lowering of the low-water level.

Cross Jetties. The erection of jetties or groynes, projecting at right angles from the banks into the channel at intervals, has often been resorted to for regulating a river, with the object of increasing its depth in wide shallow reaches. Experience, however, has shown that cross jetties, if placed far enough apart to render the system economical, do not generally deepen and regulate the channel in a satisfactory manner. The channel tends to adopt a circuitous course between the jetties, and whilst becoming deeper opposite the ends of the jetties, it shoals in the intermediate spaces.

Longitudinal Training Banks. The regulation of a channel is more effectually accomplished by longitudinal embankments, following the course of the river, and fixing precisely the widths of the low-water channel along the portions which require training. In fact, these training works furnish stable artificial banks, affording a suitable width of channel, in place of the natural banks where subject to erosion. These banks are ordinarily formed of continuous mounds of rubble stone, or chalk, protected if necessary on the face by pitching or concrete, and secured sometimes at the toe from undermining by stakes or piles. Where solid materials are not easily procurable or are costly, fascine mattresses, weighted with stones or clay, are advantageously employed. In some cases, where the alluvial matter is fairly light, floating bundles of brushwood, or fascines, attached to a pole or cord, resembling large weeds growing from the bed of the river, have been successfully used for training a river, so as to produce scour in one part and deposit in another, as for instance on the Missouri[1] and the Isar[2].

Where islands, resulting from the accumulation of deposit in mid-channel, or from a divergence of the current, split the channel in two, the two branches do not afford the same capabilities for navigation as the undivided channel above and below. Accordingly, it is sometimes necessary to concentrate all the discharge during the low-water stage of the river into one of the channels, by barring the other one up to a little above low-water level, so that a better depth may be secured in the navigable channel when the river is low, whilst leaving the secondary channel available for the passage of floods.

Protecting and Easing Bends. As the main current of a river is directed against the concave bank at a bend, the bank is subject to constant erosion (Plate 3, Fig. 1); and the sinuosity of a river through alluvial plains tends always to increase, to the detriment of navigation, till at last sometimes the loop becomes so tortuous that the river in flood-time opens out for itself a more direct channel if the intervening land is flat. To avoid an undue increase in curvature, a concave soft bank must be protected by fascines, stakes, or stones ; or the current may be kept off from the bank by short projecting spurs of stone or brushwood at a low level, a system possessing the further advantage of straightening the low-water channel, leading to a reduction of the excessive depth near the concave bank, and a corresponding diminution of the shoal at the change of curvature (Plate 3, Fig. 6). Where the curvature is already excessive, it may be eased by a flatter training bank in front of the concave bank (Plate 3,

---

[1] 'Report of the Chief of Engineers, U. S. A., 1879,' part 2, p. 1051.
[2] 'Zeitschrift für Bauwesen,' 1886, p. 515; and 'Wochenschrift des Österreichischen Ingenieur und Architekten Vereines,' 1888, p. 74.

Figs. 2, 4, and 5); but such works are liable to be costly, owing to the depth and scour close to the bank; and the width of the channel at a bend must not be reduced without a corresponding modification of the channel between the bends, which must be narrower than at the bends to maintain the depth where the current is shifting over from one bank to the other (Plate 3, Figs. 3, 4, and 6).

Straight Cuts. Sometimes it is expedient to do away with tortuous bends in a river by cutting a direct channel where the current is gentle, and thus restore the river to a primitive condition. The channel is thus rendered more convenient for navigation, whilst its length is reduced; and the increased fall ensures the maintenance of the depth of the new cut. The channel, however, should be excavated to the full size before the diversion of the river into it is effected, as its enlargement by scour would only result in the deposit of the eroded material in the channel below the cut. The banks of the cut at its upper end should be protected from erosion, and also the banks of the river immediately below the cut, as any irregularities in the direction of the channel just above or below the straight portion would lead to the development of fresh sinuosities, and eventually obliterate the benefit resulting from the rectification of the channel. Where the current is rapid, straight cuts are inexpedient for the purposes of navigation, as they augment the difficulties of the up-stream navigation by increasing the velocity of the current.

Submerged Cross Dykes. With whatever care a river may be trained, so as to secure a width suitable to the fall and the desired depth, the bottom of the river always exhibits variations in depth, resulting from the changes of the scour of the current in a winding course, from differences in the constitution of the bed, and from irregularities in fall. To obtain uniformity of depth, it is, accordingly, necessary to protect the bed from undue erosion, just as the sides of the channel are protected by training banks to secure uniformity of width. This may be accomplished by depositing dykes of rubble, or mattress sills, across the channel in the deep places, but kept below the navigable depth in the main channel (Plate 3, Fig. 5). These submerged dykes should project a little up-stream from the two banks, and dip down towards the centre of the channel, so that whilst regulating the current, they may direct it towards the centre of the river. The undue lowering of the water-level by the increased discharge in the deep parts of the channel is, moreover, prevented by these submerged works ; and the fall at a rapid is reduced by thus regulating the eroded hollow at the foot of the rapid. This method of regulating the bed, as well as the banks of a river, has been adopted for many years on rivers in Germany, and has been more recently applied to the Rhone below Lyons.

Influence of the Descent of Detritus. It is evident that the detritus brought down by rivers seriously enhances the difficulties attending their improvement ; and these difficulties largely depend upon the amount, density, and size of the detritus brought down in proportion to the discharge, and also upon the fall of a river and its variations. If the velocity of a river was uniform, or was only reduced in its descent in proportion as the facility of the transportation of the detritus is increased by its reduction in size by attrition, a regular accumulation of this alluvial matter would only be experienced at the outlet of the river. Any material reduction in velocity, however, such as is often met with, especially in torrential rivers, leads to the arrest of some of the detritus and the formation of a shoal, which cannot be coped with by dredging, owing to the abundance of the supply in many cases and its continuity. Any partial regulation of the channel merely shifts the position of the shoal ; whilst the embankment of a river, to obtain the increased scour of the flood waters for deepening the channel, or in order to prevent inundations, cuts off the lateral areas which served for the deposit of sediment, so that a larger volume of solid matter is carried down to be deposited in the channel, or to accumulate at the outlet. Accordingly, regulation works in rivers carrying down considerable quantities of detritus, must not aim at local, isolated improvements in depth, but must be carried out as a complete scheme, with a view to the uniform discharge of the detritus, by securing, as far as practicable, uniformity in velocity. Adjacent plains also, serving as natural lateral reservoirs for the deposit of detritus in flood-time, should not be shut off from the river if possible, especially where a river descending from the hills into a plain experiences an abrupt reduction in fall.

The injury caused to rivers, flowing through plains, by the descent of detritus resulting from the disintegration of mountains, has in some special cases been arrested near its source by placing a dam at a suitable place across the valley of a mountain stream, behind which the descending detritus accumulates, another dam being erected at a fresh spot as soon as the space behind the previous one has become filled up. This system, which has been occasionally resorted to in mountainous districts in Europe, has been also employed in California to arrest the devastation of alluvial lands, occasioned by the carrying down by mountain torrents of large quantities of débris left by the extraction of gold from the mines in the hills, which filled up the channels of the rivers in the plains at the base of these mountains, wasted the flat lands by deposits of gravel, and produced extensive floods [1]. The great volume of detritus, however, brought down generally from the numerous mountain sources of a large river, cannot in practice be dealt with in this way, owing to the sources being often under a separate jurisdiction, and sometimes in a different country to the main river, and also owing to the cost the works would involve, the large spaces needed for storing up the detritus, and the injury to the valleys such vast and constantly increasing accumulations

[1] 'Report of the Chief of Engineers, U. S. A., 1881,' part 3, p. 2485.

would entail. Nevertheless, the supply of detritus may be somewhat checked by stringent enactments against throwing débris into streams, or placing it so close to the banks as to be carried down in flood-time; and it may be reduced by protecting the mountain slopes from disintegration by preventing their denudation from reckless clearing, and by encouraging the growth of trees and vegetation in exposed parts. This latter course not only diminishes the supply of detritus, but also reduces the power of the torrents to carry it down, by equalizing their flow.

Remarks on the Regulation of Rivers. Though regulation works are valuable in many cases in fixing and deepening the channel of rivers, and in equalizing their flow, this system by itself is only suitable for obtaining an adequate navigable depth in large rivers with a tolerably uniform flow. If the discharge of a river, and consequently its depth, becomes very small in the dry season, it may be impossible to secure a sufficient depth, during that period, by restricting the width of its low-water channel within reasonable limits. Moreover, the removal of shoals, by facilitating the discharge, lowers to some extent the low-water level which has been raised by the contraction of the channel, and consequently reduces the gain in depth which might otherwise be anticipated. To obtain the most satisfactory results, the regulation of the sides of a river should be accompanied by a regulation of its bed, so as to secure uniformity in depth and slope as well as in width.

The large rivers of North America have been regulated in many places, where deficient in depth, with successful results; and some of the large rivers of Germany, and also the Danube, have been improved for navigation by systematic regulation works. The Rhone, however, is the only river of France that has been improved by this method alone, which indeed has been only rendered practicable in this instance by the different seasons of the floods of its tributaries, affording it a comparatively regular flow. This system, accordingly, has a somewhat limited application ; and other means have to be resorted to for improving the navigable capabilities of rivers having smaller, or more variable discharges.

Instances of Regulation Works. The Rhine has a very moderate fall between Plittersdorf, about 29 miles below Strassburg, and the sea, varying on the average from 1 in 4,300 down to 1 in 10,000 near its outlet, with the exception of the portion between Bingen and St. Goar, where the fall is increased to 1 in 2,000, and the current is very rapid in places in a narrow rocky channel. The navigable condition of the river has been gradually improved by extensive regulation works below Strassburg, reducing its low-water channel in width, so that it has a minimum depth of 3 feet below Strassburg, increasing to 6½ feet above Mannheim, which continues to Bingen (Plate 3, Fig. 2). At the rapid part of the river, between Bingen and St. Goar, the minimum navigable depth barely attains 5 feet, in spite of the removal of rocky obstructions ; but below St. Goar the minimum depth reaches 7¼ feet, and continues the same down to Coblenz and on to Cologne, below which a depth of about 8 feet is obtained, increasing somewhat in Holland. This large, unimpeded waterway, extending a long distance inland, though with a very moderate minimum depth, provides the route for a very large inland trade. reaching altogether about ten million tons annually in Germany alone, and has enabled a very flourishing port to be established at Mannheim, 352 miles from the sea, with a traffic of nearly two million tons annually.

The Elbe has been trained by a combined system of longitudinal banks and dipping cross jetties, with submerged dykes at the end of some of the cross jetties to protect them, and also in front of the longitudinal banks in some places to reduce the excessive depth at their toe (Plate 3, Fig. 3) [1].

---

[1] 'Handbuch der Ingenieurwissenschaften, Band III, Der Wasserbau,' L. Franzius and Ed. Sonne, plate 7.

These works have provided a depth, at the ordinary water-level, of 5 feet at the German frontier, down to $6\frac{1}{2}$ feet at Hamburg, a distance of 316 miles; but the minimum depth of 3 feet at the lowest water-level has been obtained with difficulty, and a further contraction of the low-water channel was found necessary. The Elbe has thereby been constituted an important waterway for inland navigation, second only to the Rhine in Germany.

The Niemen also has been deepened $1\frac{1}{8}$ to $1\frac{2}{3}$ feet by similar training works, increasing its navigable depth to $5\frac{3}{4}$ feet at mean low-water (Plate 3, Fig. 4).

The regulation of the Rhone between Lyons and the sea, along a length of 201 miles, was commenced in 1860, and has been continued up to the present time with important modifications[1]. The Rhone is not navigable above Lyons at the low-water stage, its bed even through the upper part of the town being strewn with shoals of gravel and stones. Even below Lyons, where its flow is augmented and regulated by the confluence of the Saône, its fall is rapid in many parts, especially for some distance below Valence, where it averages about 1 in 1,280, and is also irregular; and its bed being readily eroded in most places, the deepest channel was frequently shifted by floods. The works first undertaken were designed to fix the main channel, and to increase the depth in the worst places, by shutting off secondary channels, and by training the river up to a mean water-level by longitudinal stone embankments, so as to lower the shoals by the scour of the concentrated current. These embankments, raised from $6\frac{1}{2}$ to 10 feet above the low-water level, had necessarily to be placed too far apart to train the low-water channel, which consequently meandered between shoals, crossing over from one concave bank to the next on the opposite side.

[1] 'The Training of Rivers, as illustrated by the Results of various Training Works,' L. F. Vernon-Harcourt, Minutes of Proceedings Institution C. E. 1894, vol. cxviii, p. 4, and plate 1, figs. 1-5.

Moreover, the high embankments made the current, when
filling the trained channel, scour deep holes along the concave
bends. The river accordingly, even in the trained portions,
though considerably improved, was not uniform in depth ;
and the removal of the shoals, by lowering the water-level
in the trained portions, increased the fall and reduced the
depth above the ends of the training works, so that the
improvements in some parts led to the appearance of fresh
obstacles at others. The positions, indeed, of the shoals and
rapids were altered ; but neither uniformity of depth nor of
fall was attained, and therefore the navigable capabilities of
the river at a low stage were not much improved by these
training works.

Since 1880, a new system has been adopted ; and the
river has been trained as a whole, instead of being dealt
with in isolated portions [1] By reducing the height of the
training banks, it has been possible to diminish the width
of the trained channel, so as to correspond more nearly
with the low-water stage, and thus prevent to some extent
the undue lowering of the water-level at low water. In
order, however, to render the fall fairly uniform, which
varied with the differences in depth, submerged dykes
projecting from the training banks slightly up stream, and
dipping down towards the centre of the channel, have
been placed in the deep places, so as to regulate the bed,
and have thus reduced the fall over the shallower parts by
increasing it in the deep places (Plate 3, Figs. 5 and 6).
These submerged dykes, moreover, besides raising the lowered
water-level, and thus increasing the available depth, possess
the advantage of directing the main current towards the
centre of the channel, and thus straighten the navigable
channel, and keep the vessels away from the concave bank
at the bends.

[1] ' De l'Amélioration des Rivières navigables à fond mobile,' M. Jacquet.
' Congrès international de l'utilisation des eaux fluviales,' Paris, 1889, p. 159.

The available minimum navigable depth of the Rhone
below Lyons, down to Arles, has been raised by these works
to $3\frac{3}{4}$ feet, and $5\frac{1}{4}$ feet below Arles, whilst the river does
not fall to its lowest level every year; and the annual traffic
has attained 236,000 tons. The trade, however, along the
river does not appear at all commensurate to the very
extensive training works which have been carried out. Undoubtedly, the rapidity of the current must always prove
a hindrance to the up-stream navigation, though various
systems of towage have been proposed for minimizing its
difficulties; and the cost of the works and their duration
would have been greatly reduced if the improved system of
training had been adopted at the outset. Nevertheless, two
serious obstacles have existed, up to the present time, to
the proper development of navigation on the Rhone. The
one is the absence of competition in conducting the navigation along the river, as the towage has hitherto been
monopolized by a single company, owning about sixteen
steamboats serving both as tugs and for carrying cargo; but
towage by means of a fixed submerged chain, wound up or
unrolled by a drum on the tug, is being tried by a new
company at Lyons, with a view of working the navigation
in a series of sections corresponding to the lengths of the
chains laid down. The other obstacle is the necessity of
transhipping the cargoes carried by water between Lyons
and Marseilles at Port St. Louis which might be obviated
by enlarging the old canal going from Arles to Bouc, and
prolonging it to Marseilles, so as to connect the Rhone
directly with Marseilles without the intervention of a sea
passage, and thus enable river-craft to go from Lyons
right into the port of Marseilles.

The above instances show that a considerable traffic can
be carried on improved inland waterways possessing very
moderate minimum depths, to which, however, the rivers only
fall for short periods, and not every year; whilst the history

of these and other inland navigations indicate that the trade quickly increases with improvements in depth and accessibility.

CANALIZATION OF RIVERS.

In rivers where the low-water discharge is small, and a considerable improvement in the minimum depth is required for navigation, it is necessary to obtain the requisite increase in depth by raising the low-water level. This system, moreover, possesses the advantages over training works, not merely of affording greater scope for improvements in the navigable channel, and of extending notable improvements in depth to moderate-sized rivers, but also of rendering it possible to obtain a further increase in depth without difficulty if circumstances make it expedient, and of making the up-stream traffic almost as easy as the down-stream.

Primitive Method of River Navigation. In old times, before any alterations had been made in the natural condition of rivers, trees felled in the forests were floated down stream, a plan still in common use in mountainous regions, and in places devoid of roads. Subsequently, the timber was formed into rafts capable of carrying down some of the produce of the upper country, where the flow of the river was adequately regular; and next small boats were attached to the rafts, better suited for the safe conveyance of goods, which were taken to pieces and used as timber on reaching their destination. In France, at the present time, 629 miles of rivers are classed as floatable. This traffic, however, can be merely carried on down stream, and only when there is sufficient water in the river to float the rafts and boats over the numerous shoals met with in their course.

Stanches for Flashing. Later on the idea was conceived of producing artificial floods, by damming up the whole discharge of the river for a certain period, till a considerable quantity of water had accumulated behind the dam, and then letting it suddenly escape. This was accomplished by means of stanches which consisted of spars, planks, or paddles, supported by the pressure of the water against a sill below and a movable beam above. These stanches were erected across a narrow place in the river ; and a man standing on a foot-bridge above was able to open or close them as required. The level of the water was raised for a considerable distance above these stanches when closed; and the mass of water being set free when the stanches were suddenly opened, the boats, which had previously collected above, were floated over the shallows below. This operation of flushing, or flashing, as it is termed, continued in general use on several French rivers till past the middle of the nineteenth century, and was also employed on the Thames and the Severn[1]. The stanches on the Severn were removed in 1842, when some improvement works were carried out ; but some stanches are still in existence on the Thames above Oxford, where they chiefly serve for keeping up the water-level in summer, and are then generally only partially opened for the passage of the few boats frequenting that part of the river.

A novel form of stanch was introduced by Mr. Poirée on the river Yonne in 1834, consisting of a series of iron frames, placed about 3¼ feet apart across the river and carrying a foot-bridge, each connected by a bar on the up-stream side, against which and a sill at the bottom, a row of long wooden spars rested, square in section, placed close together nearly vertical, thereby closing the channel[2] (Plate 4, Fig. 7). The slender spars, or needles (*aiguilles*) as they are termed in France, were readily raised successively or lowered by a man from the foot-bridge thus opening or closing the stanch. At that period the system of navigation by aid of flashes was employed on the Upper Seine, the Yonne, and other

---

[1] Minutes of Proceedings Institution C. E., vol. iv. p. 111.
[2] Annales des Ponts et Chaussées, 1839 (1), p. 238.

French rivers; and it is continued at the present time in France on the Upper Yonne and some small floatable rivers. Flashing is very inconvenient for up-stream navigation ; and the danger experienced by descending vessels in the rapid passage at the stanch, led eventually to the addition of a lock alongside in some cases, so that vessels might avoid passing through the stanch whilst making use of the flashing for the rest of their journey. This system, moreover, only affords an intermittent navigation for the down-stream traffic; and therefore this method of river navigation compared unfavourably with the continuous navigation afforded by canals, which were extensively constructed during the latter half of the eighteenth century, and the earlier part of the nineteenth century.

Locks, Weirs, and Level Reaches. Locks, which are supposed to have been first employed on some canals in Italy in the fifteenth century, were eventually introduced on rivers, for the purpose of securing a continuous navigation with an improved navigable depth. Instead of increasing the depth of the river by lowering its bed, the level of the water is raised. The river is transformed, by dams or weirs, into a series of nearly level reaches connected by locks whose chambers, being successively filled with water and emptied, serve to raise or lower a vessel from one reach to the next. By this means the requisite navigable depth is obtained with little excavation, except at the portions of the river just below the locks, and at old fords ; but the banks have sometimes to be raised for a certain distance above the locks and weirs to prevent the river overflowing its banks when its water-level is raised. The position of the locks, and the length of the reaches, are regulated by the slope of the valley ; the fall at a lock generally ranging between five and thirteen feet. The lock is usually placed where a river is divided into two channels with an island between, the lock being situated in one channel, and a weir, keeping up the water to the level necessary for navigation, across the other channel (Plate 3, Fig. 7, and Plate 4, Fig. 4). If no natural division of the river occurs at a point sufficiently close to the place where a change in the water-level is desirable, a site is selected where the river makes a bend, a new straight channel is excavated across the bend, and the lock is constructed in this channel (Plate 4, Fig. 5). The ordinary discharge of the river follows the old channel, and flows over the weir placed in the bend. This latter arrangement, though necessitating more excavation, is preferable, as it interferes less with the original channel.

This canalization of a river, converting it into a fairly still-water navigation, except during floods, greatly facilitates the up-stream traffic, as the fall of the river is mainly concentrated at the weirs under ordinary conditions of flow, and the current of the discharge is reduced by the deepened waterway. Moreover, the level of the river, and consequently the depth, does not fall below the amount regulated by the height of the weir in the driest weather, unless the expenditure of water in lockage, evaporation, and leakage, exceeds the discharge, which is very rarely the case.

Instances of Canalized Rivers. The extent to which the minimum navigable depth has been improved by canalization is indicated by comparing the former low-water lines on the Upper and Lower Seine, and the Main, with the present water-levels retained by weirs (Plate 4, Figs. 1, 2, and 6). This increase in depth, moreover, upon the Seine has been attained by successive stages, by raising, and in some cases reconstructing the weirs. The canalization of the Upper Seine was commenced in 1860, with the object of obtaining a minimum navigable depth of $5\frac{1}{4}$ feet; but by works undertaken in 1878, the depth has now been raised to $6\frac{1}{2}$ feet, ten of the weirs having been raised, and the two top reaches deepened [1] (Plate 4, Fig. 1). Improvement works were com-

---

[1] 'The River Seine.' L. F. Vernon-Harcourt. Minutes of Proceedings Institution C. E., 1886, vol. lxxxiv, p. 228.

menced on the Lower Seine below Paris at the beginning of
the nineteenth century; but as the dredging of the shoals
lowered the river, little advance was made in its navigable
capabilities, till its canalization was undertaken in 1838, and
completed in 1866, raising its minimum navigable depth to
$5\frac{1}{4}$ feet[1]. Directly these works were completed, supplementary works were determined upon for increasing the depth
to $6\frac{1}{2}$ feet; and in 1878, before the termination of these
works, further works were authorized, which have increased
the minimum depth to $10\frac{1}{2}$ feet by raising the water-level at
the weirs, an additional lock and weir having also been constructed, and some shoals lowered by dredging. These works
enable vessels of 800 to 1,000 tons, and with 9 feet 10 inches
draught, to navigate the Seine between Paris and Rouen at
all times, except when floods rise above the navigable limit.

The canalized Seine between Montereau and Martot, at the
limit of the tidal river, a distance of 197 miles, has an inland
navigation trade second only in France to that of the northern
canals which bring the produce of the coal-fields of the north
of France and Belgium to Paris. The traffic, however, varies
in the different sections according to their position in regard
to Paris, as the feeders of the traffic converging mainly to
Paris join the river, and not in proportion to the navigable
depth; for the traffic on the Upper Seine in 1892 amounted
to 1,114,000 tons between Montereau and Corbeil, and reached
2,295,000 tons between Corbeil and Paris; whereas, on the
Lower Seine, the minimum traffic of 1,235,000 tons was
on the portion between Rouen and the mouth of the Oise,
and the maximum of 3,337,000 tons in the next section above,
between the mouth of the Oise and La Briche, where the
St. Denis Canal diverts a portion of the up-stream traffic by
affording a more direct route to Paris. The river traffic between
Corbeil and Paris furnishes another illustration of the large

---

[1] 'The River Seine.' L. F. Vernon-Harcourt. Minutes of Proceedings Institution
C. E., 1886, vol. lxxxiv, p. 225.

trade that can be carried along a wide waterway of very moderate minimum depth. Moreover, though the traffic on the Lower Seine, with a minimum depth of 10½ feet, is less in one section than in the most frequented section on the Upper Seine, with a depth of only 6½ feet, the deeper waterway enables the transport to be effected more economically, and could accommodate a much larger trade.

The Thames and some other English rivers were gradually canalized at a considerably earlier period than the Seine; whilst a large work of canalization was carried out on the Main from its junction with the Rhine up to Frankfort, a distance of 22 miles, as recently as 1883–86 (Plate 4, Figs. 5, and 6)[1]. The Main had been previously trained by longitudinal embankments on the left bank, and cross jetties along the right bank; but the minimum navigable depth, at the low stage of the river, obtained by these works did not reach the contemplated 3 feet; and the annual river traffic between the Rhine and Frankfort was only about 12,000 tons. The canalization, however, aided by dredging, increased the minimum depth to 6½ feet, enabling vessels of 700 to 1,000 tons to go up to Frankfort; and the traffic rose in 1887 to 300,000 tons. The remarkable success of these works led to a decision, in 1889, to increase the depth to 8¼ feet, which is being carried out by dredging near the upper end of each reach; and the traffic attained 709,000 tons in 1892.

Remarks on Canalization of Rivers. The above examples of canalization illustrate the increasing difficulty which is generally experienced in obtaining an adequate navigable depth in ascending a river, or in its tributaries; for whereas a navigable depth of 10½ feet has been obtained in the Lower Seine by locks and weirs placed on the average about 13½ miles apart, the depth of 6½ feet on the Upper Seine has only

---

[1] 'Canal, River, and other Works in France, Belgium, and Germany,' L. F. Vernon-Harcourt, Minutes of Proceedings Institution C. E., 1889, vol. xcvi, p. 189.

been effected by locks and weirs placed at average intervals of only 4¾ miles (Plate 4, Figs. 1 and 2). Whilst also the Rhine has been adequately improved by training works up to Mannheim, 45 miles above the confluence of the Main, training works proved inadequate in this latter river; and locks and weirs have had to be placed in it, only 4¾ miles apart on the average (Plate 4, Fig. 6). The length of the weirs is, indeed, less in the upper part of the river, owing to the reduced width of the river ; but this affords a very inadequate compensation for the increased number of locks and weirs required, so that canalization becomes more costly in the upper parts of a river valley on account of the increasing slope.

Canalization is most useful where, owing to unfavourable conditions, training works are unable to realize an adequate improvement in a river ; and it enables navigation to be extended to rivers, and to the upper portions of rivers and their tributaries, which training works would fail to ameliorate. Canalization, moreover, affords a much greater certainty of obtaining a definite increase in depth than training works, which, in deepening a river, are liable also to modify its water-level ; and though somewhat impeding the down-stream traffic by reducing the current and imposing a delay at the locks, canalization renders it more safe, and greatly facilitates the traffic up-stream.

CHAPTER IV.

DREDGING AND EXCAVATING.

Definition of Terms. *Dredgers:*—Bag and Spoon; Bucket-Ladder Dredger, Stationary and Hopper, details, cost of machines, and cost of Dredging; Dipper Bucket Dredger; Grab Bucket Dredger, varieties, and capabilities; Sand-Pump Dredger, instances, capacity for work and cost of Dredging; Eroding Machines; Remarks on Dredgers; Removal of Rock under Water. *Discharge of Dredged Materials*—through a Hopper; by a Chain of Buckets; direct into Wagons; through Long Shoot; by Pump and Floating Tubes. *Excavators:*—Bucket-Ladder Excavator, method of work, cost, capabilities; Steam Navvy, varieties, cost, capacity for work; Grab Excavator; Remarks on Excavators.

DREDGING is the term applied to excavations carried on under water ; whilst excavating refers to similar operations conducted wholly on land. Extensive excavations are carried out by means of large machines, worked by steam, which are called dredgers when fitted on barges for removing material under water, and excavators or steam navvies, when running on wheels and excavating on land. Dredging is employed for deepening the channels of rivers, removing obstructions and shoals, lowering bars at the mouths of rivers, and enlarging canals. Excavators are only occasionally used in river works, for forming a new cut through dry land ; but they constitute a very important part of the plant in the construction of large canals. Excavators are made on the model of some of the principal types of dredgers, being merely adapted for working on land. Dredging, however, is fre-

quently carried out more economically than excavating, owing
to the comparatively low cost of transport by water.

DREDGERS.

Bag and Spoon. The simplest form of dredging machine
is the bag and spoon, which consists of a leather or canvas
bag, having a circular ring of iron round its mouth, fastened
at the end of a long pole (Fig. 10). A
man standing in a barge holds the upper
end of the pole, and guides it; whilst
another man drags the bag along the
bottom by means of a chain fastened to
the lower end of the pole, and winding
round a crab placed at the far end of the
barge. The mud or gravel is scooped
up by the flat lower part of the ring, and
enters the bag, which is drawn up by the
chain; and the contents of the bag are
deposited in the barge. The chain is
then slackened, the bag is lowered to
the bottom, and the process repeated.
This primitive plan is still employed for
small operations in tolerably shallow
water, and is constantly used for getting
gravel from the bed of the Thames.

Bag and Spoon.
Fig. 10.

The aquamotrice is a modification of the bag and spoon, in
which an iron scoop, or small bucket, is hinged to its handle,
so that it can be turned over for discharging on releasing
a catch; and the dragging of the scoop along the bottom is
effected by aid of the current of the river [1]. The barge
carrying the scoop is moored over the site to be dredged; and

[1] Annales des Ponts et Chaussées, 1874 (2), p. 188.

paddle wheels, on each side of the barge in front, in revolving under the action of the current, wind up a chain passing over pulleys and attached to the scoop, and thus drag it along the bottom, and raise it at the end of its course to be discharged. Directly the scoop has discharged its load, the windlass winding up the chain is put out of gear, till a sufficient length of chain has been released for the scoop to return to the starting point again, when the operation is repeated. Dredging in gravel and shingle was accomplished on the Garonne, at Agen, with this machine, raising 65 cubic yards of gravel per day of 12 hours, at a cost of $1\frac{1}{2}d.$ per cubic yard, including maintenance and depreciation.

Bucket-Ladder Dredger. The type of machine most commonly employed for extensive dredging operations is the bucket-ladder dredger (Plate 2, Fig. 1)[1]. It consists of a continuous row of buckets fastened to two parallel endless chains running on rollers and revolving round two tumblers at the ends of a long girder, one end of which is fixed to staging on the deck of the vessel, and the other end can be lowered as required by chains attached to it, so that the lowest bucket may come in contact with the bottom to be dredged. The buckets, passing in succession along the bottom in a horizontal position, scoop up the material, and revolving gradually, rise to the surface in a more upright position, and after passing round the upper tumbler, turn bottom upwards, and deposit their contents, through a shoot, into a well in the hold of the vessel, or into a barge alongside. The dredger is moved slowly either sideways or forwards during the operation, so that each bucket may have material to dredge. The steam-engine on the dredger moves the buckets by causing the upper tumbler to revolve.

Usually the ladder of buckets is situated in a well along the centre line of the vessel, and is inclined at an angle which

---

[1] The illustration of a bucket-ladder hopper dredger has been taken from a drawing furnished me by Messrs. W. Simons and Co. of Renfrew.

varies according to the depth to which the buckets are lowered. Sometimes the ladder is situated alongside the vessel; and occasionally a dredger carries two ladders of buckets, one on each side of the vessel.

The staging carrying the ladder is sometimes provided with traversing gear, enabling the ladder of buckets to be moved forwards so as to project in front of the bow of the vessel, and thus allow the dredger to cut a channel for itself into a shoal in advance.

Two distinct forms of bucket-ladder dredgers are employed. namely, stationary dredgers, and hopper dredgers. Stationary dredgers are so called because they remain stationary at the spot where they are dredging, and discharge the dredged material into barges alongside, which carry away the material to the place of deposit ; but these dredgers are now generally made self-propelling, so as to obviate the necessity of using a tug whenever the position of the dredger has to be changed. Hopper dredgers are provided with a well in the centre of the vessel, into which the dredged material is discharged ; and directly the well has been filled, the vessel ceases dredging, and itself carries the load to the place of deposit. The bottom of the well is formed by hinged doors which open downwards when released, and discharge the contents of the well into the water.   This arrangement of a well with a movable bottom is termed a hopper; and by this means a dredger and a depositing hopper barge are combined in a single vessel.

The stationary dredger possesses the advantage of being able to remain continuously at work so long as there is sufficient water to float it, provided it is served by an adequate number of barges, depending upon its rate of dredging and the distance of the place of deposit. The hopper dredger has, on the contrary, to work discontinuously, its dredging being stopped during the period required for leaving its moorings, steaming down to the depositing ground, releasing its doors and closing them again, returning, and picking up its moorings; and during these operations, the dredging machinery remains idle, and merely increases the load to be transported. A stationary dredger with hopper barges should, accordingly, be employed when the amount to be dredged is large, and in fairly thick layers, when there is ample space in the channel for the dredger and its barges alongside, when the place of deposit is distant, and when the cost of an ample dredging plant is moderate in proportion to the total cost of the dredging to be carried out. When, however, the shoals are scattered, the thickness of material to be removed is small, and the total amount to be dredged is moderate, when the depositing ground is not far off, when subsequent maintenance has to be provided for, and when the space available for the dredging plant to work in is limited, it is expedient to resort to a hopper dredger. This type of dredger, though executing the work more slowly than a dredger with barges, occupies less room, is handled more easily, and with a smaller crew, is less costly, and is better suited for the comparatively moderate amount of dredging requisite for maintenance. Moreover, it can be supplemented when desirable by an attendant barge, which it can load and take in tow; but, on the other hand, its increased draught as its hopper is loaded, reduces its period of work on a falling tide in a shallow river. The very extensive dredging operations carried out during many years on the Tyne, the Clyde, and the Tees (Plate 8, Figs. 2, 5, and 8), have been effected by bucket-ladder dredgers with attendant steam hopper barges, or hopper barges towed out by tugs; whilst many smaller improvements in rivers and harbours, of which the approach channel to Belfast is a notable instance, have been accomplished in recent years by bucket-ladder hopper dredgers.

The dredger is gradually shifted forwards or backwards, and sideways whilst dredging, by hauling with steam winches, placed fore and aft, on chains attached to buoyed moorings, generally six in number, laid ahead, astern, and on each side of the vessel, so that the chain of buckets may successively traverse the area comprised within the moorings, and lower it to a uniform depth.   A dredger is often provided with twin screws, so that it may be navigated more easily in narrow or tortuous channels, and may be more quickly placed in position for working.   One of the largest dredgers hitherto constructed, which, being built for the port of Bristol, has to navigate the narrow River Avon, is provided with two propellers both fore and aft, so that it can steam in either direction without having to turn.   A similar arrangement has been adopted in the large hopper dredger constructed for dredging a channel in the Mersey at the entrance to the Manchester Ship-Canal (Plate 2, Fig. 1); and the ladder-well has been carried through the stern, instead of through the bow, in order to obtain greater seaworthiness and speed by leaving the bow undivided.

The buckets, tumblers, links, and pins, are now made of steel; and the bottom of the vessel is advantageously made of steel if liable to have to ground at low water.   The Bristol dredger was made wholly of steel[1].   Cast-steel buckets have been recently introduced, which, being formed in one piece, are stronger than riveted buckets; and manganese steel has proved a very durable material for the pins which are subjected to great wear.   The buckets vary in size according to the size of the dredger; and they should also vary according to the nature of the material, a larger bucket being more suitable for sand and soft material than for stiff clay and hard material.   A dredger, however, has often to work in various kinds of soil, for which purpose two sizes of buckets are occasionally provided.   The largest sized bucket has a capacity of about 23 cubic feet; the buckets of the Bristol dredger have a capacity of 17½ cubic feet, those of one of the Clyde dredgers 15 cubic feet, of the Swansea dredger 13 cubic feet, and of the Tees dredger 9 cubic feet; whilst the buckets of

[1] Minutes of Proceedings Institution C. E., vol. cvii, p. 316.

small dredgers are still less. The buckets should taper out from the bottom towards the mouth, so as to facilitate the discharge of their contents. They used generally to be made with holes in their sides, to let the water escape; but now the holes are often omitted, as the water, though adding to the weight lifted, is useful in facilitating the discharge of the materials from the buckets, and down the shoot. In dredging stiff or hard material, ripping claws are sometimes added to the chain of buckets at intervals to loosen the soil; but the buckets with steel lips, in the stronger machines, are capable of dredging boulder clay, soft sandstone, and rock in layers. The rate of working varies with the nature of the material; and as the buckets are often only partially filled, it is safer to estimate the capabilities of a dredger by the amount actually delivered into a hopper in a given time, than to calculate it from the speed and capacity of the buckets.

Bucket-ladder dredgers are commonly built capable of lifting from 200 to 1,000 tons of free soil in an hour under favourable conditions, with maxima depths of working of from 25 to 35 feet. Hopper dredgers are provided with hoppers of from 250 to 1,300 tons capacity, according to the size of the dredger. The cost of these dredgers ranges generally between £8,000 and £30,000. Thus the four stationary dredgers employed for several years past on the Clyde, cost from £6,650 to £17,650[1], and have dredged in depths of from 28½ to 34 feet; whilst one of these raised an average quantity of nearly 200 cubic yards, or about 250 tons per hour in 1890-91. The four dredgers working in the Tees cost from £10,130 to £18,100[2]. The Swansea stationary dredger with two ladders of buckets, raising on the average 320 tons per hour, and a maximum of about 600 tons, cost £27,500. The Bristol hopper dredger, with a hopper capacity of 1,000 tons,

[1] 'The River Clyde.' James Deas. Minutes of Proceedings Institution C. E., 1873, vol. xxxvi, p. 160.

[2] 'Dredgers and Dredging on the Tees,' John Fowler. Minutes of Proceedings Institution C. E., 1884, vol. lxxv, p. 239.

and capable of dredging a maximum quantity of 800 tons in an hour, cost £30,000. Steam hopper barges with hopper capacity of 200 to 500 tons, cost from about £4,000 to £7,000 ; and non-propelling hopper barges having similar capacities, together with a steam tug, approximate to the same cost.

The speed of a hopper dredger is of considerable importance, and also its draught, especially when working in a tideway, for on these, and the rate of dredging, depends the number of trips that can be accomplished in a day. The speed of these vessels ranges between 7 and 10 miles an hour ; the draught varies with the size, from about $6\frac{1}{2}$ to $10\frac{1}{2}$ feet when light, and $10\frac{1}{2}$ to $14\frac{1}{2}$ feet loaded. Two sets of engines work the twin screws ; and one, or both if necessary, can be used for dredging. When dredging plant has to be sent to a distant country where barges are not available, the hopper dredger possesses the important advantage of being self-contained, and a seaworthy vessel ; whereas the transport of barges with a stationary dredger presents difficulties.

The cost of dredging with bucket-ladder dredgers depends on the nature and compactness of the material, the concentration and amount of the work, the depth of water and the velocity of the currents, the period during which dredging is practicable and the exposure of the site, the distance of the place of deposit, the cost of wages and coals, and the efficiency of the dredging plant. All these conditions may vary with the locality, so that no definite price can be arrived at ; whilst the problem is complicated by the fact that many records of the cost of dredging only take into account the actual working expenses, without making allowance for repairs, interest on the capital cost of the plant, and its depreciation. Nevertheless, the actual expenditure in various dredging works, reckoned per ton or per cubic yard removed, affords a sort of basis on which to found a rough estimate of the cost of dredging under similar conditions. The price is generally given per ton, as this is a measure of the actual work accomplished, and as the contents of the hoppers are reckoned in tons; but the improvement effected in a channel depends upon the volume removed, and therefore should strictly be estimated in cubic yards measured in place. The work done, however, is often more than would appear from comparing the cross-sections before and after the dredging, owing to the silting which frequently takes place during the deepening of a river; but it is desirable to know the relation between a ton and a cubic yard of material, wherever dredging is carried out. On the Clyde, 1¼ tons of dredging are considered to occupy a cubic yard; whereas on the Tyne, 1½ tons are supposed to be the equivalent.

The actual working expenses of dredging with bucket-ladder dredgers and depositing, are sometimes less than 2*d.* per ton; whilst on the Tyne, including maintenance and repairs, the cost of dredging has varied from 2½*d.* to 8½*d.*, and has averaged 3·39*d.*, the material being deposited in the sea at an average distance of 11 miles from the work[1]. The Bristol dredger has raised mud, sand, and sandstone, and deposited it 10 miles off, at a cost of 5*d.* per ton, exclusive of depreciation. It is, however, far more satisfactory when the price comprises every item of expenditure, including repairs, interest on cost of plant, and depreciation. On the Clyde, the total cost of dredging boulder clay amounted, in 1872, to 2*s.* per ton, and in dredging sand and silt 6½*d.* per ton, which in 1880 had been reduced to 4⅔*d.* per ton, the distance of the depositing ground being from 9 to 27 miles. At Port Glasgow, the total cost of dredging and of carrying the dredgings 7½ miles, has been 3¾*d.* per ton. The dredging of the direct approach channel to Belfast through sand and clay, and the conveyance of the material 10 miles, cost 3·21*d.* per ton; whilst the deepening of the Tees in sand, silt, clay, stones, and boulders, and carrying the dredgings about 9 miles on the average, cost 3½*d.* per ton. The cost of dredging a channel

[1] Minutes of Proceedings Institution C. E., vol. lxxxix, p. 100.

to deep water in Swansea harbour [1], through sand, clay, gravel, and boulders, and depositing the materials 7 miles off, reached nearly 7*d.* per ton; whereas at Carlingford Lough, the removal of hard clay and boulders cost 1*s.* 5*d.* per ton [2]. The inexpediency generally of getting dredging done by contract is indicated by a tender for the work at Carlingford Lough having amounted to 1*s.* 11*d.* per ton, and by the dredging by contract in Boulogne harbour having cost 10*d* per ton, for mud, sand, and stones, and 2*s.* 2*d.* in compact schistose clay, when the place of deposit was only 2 miles distant [3]. Further instances of the average cost of dredging in mud, sand, and clay, with hopper dredgers, are 2.06*d.* per ton at the Port of Blyth, the material being conveyed 2 miles; 3*d.* at Bombay, conveyed 5 miles; and 4*d.* on the east coast of Scotland, conveyed 18 miles [4].

Dipper Bucket Dredger. The dipper dredger resembles in principle the bag and spoon, but in a greatly enlarged form: and it is worked by a steam-engine. The dredging is effected by a single large cylindrical bucket, made of iron or steel, with a projecting lip, fastened to the end of a long beam carried by a revolving derrick projecting over the extremity of a barge, and guided by chains. The scope of action of the bucket depends on the length of its beam, and the projection of the derrick; and the whole area within the radius commanded by the machine can be dredged before any shifting of the barge is required. Buckets have been made large enough to hold 2½ cubic yards of material, and the largest of these dredgers have raised 1,200 cubic yards of silt in a day of ten hours. The bucket is provided with a hinged bottom opening downwards, by means of which the contents of the bucket are readily discharged into a barge alongside; and it is sometimes provided with pointed teeth projecting in front of the

---

[1] Minutes of Proceedings Institution C.E., vol. ciii, p. 354.
[2] Ibid. vol. xliv, p. 135.   [3] Ibid. vol. lxxx, p. 263.
[4] 'On Dredging and Dredge Plant,' A. C. Schonberg, p. 20, Manchester Inland Navigation Congress, 1890.

lip, to break up compact layers. The method of working of this machine is precisely analogous to that of the steam navvy which works on land (Plate 2, Fig. 9). This type of dredger is suitable for dredging materials of any quality from silt to loose rock, in depths ranging between 5 and 35 feet.

Grab Bucket Dredger. This form of dredger consists essentially of an iron or steel semicylindrical, or hemispherical bucket, opening at the bottom into two or three sections (Plate 2, Figs. 2, 4, and 5). The bucket is suspended by chains from the jib of a crane, or by a spear from a derrick; and it is lowered, with its jaws kept open, on to the bottom to be dredged, penetrating the surface by its impact and weight. In being raised, the jaws are released, and tend to close by the weight of the sections aided by lever bars attached to them; and the sections coming together, excavate the material into which they have penetrated, and raise it in the closed bucket. The material lifted is readily discharged from the bucket, when in position, by slackening the lever chains and opening the jaws again, leaving the bucket ready for another descent. The jaws are sometimes furnished with teeth to facilitate the penetration of the bucket into dense material, for which purpose the three- or four-bladed hemispherical bucket is also suitable; and occasionally rows of claws are substituted for the sides of the bucket, to disintegrate hard material, and to lift boulders, draw up piles, and remove large obstacles. The buckets are made of various sizes according to the nature of the work, ranging in capacity from about 6 cubic feet, up to nearly 5 cubic yards.

The grab dredger is best suited for dredging silt, soft sand, and small gravel, or other loose material, and for raising large loose boulders, blasted rock, or rough débris such as trunks of trees; and it is able to dredge at any depth, and in exposed situations, and is specially well adapted for dredging in docks, locks, and other places where the available spáce is limited. The jaws of the bucket do not readily penetrate compact soils and stiff clay; but this defect may to some extent be overcome by the addition of claws, by weighting the bucket, or by pressing it into the soil by means of a spear or by hydraulic power (Plate 2, Fig. 6), and thus rendering it capable of dealing with these materials, where other forms of dredgers are not available, and where the work is small and scattered. When working in materials of various sizes, stones or other hard substances are liable to be caught sometimes between the jaws of the bucket when closing, so that an aperture is left, through which the finer materials drop out of the bucket whilst it is being raised.

These grab dredgers are generally carried on a barge provided with a central hopper[1], as the machinery is comparatively light and occupies little space, and as these dredgers often have to work in confined spaces (Plate 2, Figs. 2 and 3). The adaptation of hydraulic power to grab dredgers increases their power of penetration, and their speed of working; and the hydraulic power is not only utilized for pushing the blades of the bucket into the soil, but also for opening and closing the bucket. The largest grab dredger consists of a steamer fitted with a central hopper of about 1,200 tons capacity, and carrying four grab buckets, each holding about 1⅓ cubic yards, worked by steam cranes, and capable of raising about 3,000 tons of mud in a day, which is used for removing the accumulations of silt from the Liverpool docks. The cost of dredging with a grab varies greatly with the nature of the material and its compactness or tenacity; but under favourable conditions, or with powerful machines, the actual working expenses of dredging have in some cases been only between 1*d.* and 2*d.* per ton, whilst the cost of the plant is comparatively small.

[1] The illustrations on Plate 2, Figs. 2, 3, and 4, have been reduced from drawings furnished me by Messrs. Priestman Brothers, of Hull.

Sand-Pump Dredger. When the bottom of the channel to be dredged is composed of sand or silt, the material is readily raised by a sand-pump or suction dredger. This machine consists of a centrifugal pump fixed on a vessel, which draws up a continuous stream of the sand or silt, mixed with water, through a pipe dipping slightly into the material to be raised, and deposits it into the hopper of the dredger, or into a barge alongside [1] (Plate 2, Fig. 8). The material settles in the hopper; and as the filling proceeds, the water rising to the top of the hopper, eventually flows overboard at the sides. To give more time for the deposit of the suspended matter as the hopper fills up, flap-doors are sometimes raised round the sides of the hopper, retaining the water temporarily above the top of the hopper, so that it may discharge its burden of silt before flowing overboard, and thus enable the hopper to be filled with deposit. The settlement of the material in the hopper has also been secured by propelling the mixture of sand and water through pipes running the whole length of the hopper, and perforated with a series of holes at the bottom, through which the sand drops down. Pure sand is the material best removed by the sand-pump dredger, for an admixture of silt reduces its mobility; whilst fine silt does not readily settle in the hopper, and a large proportion is carried overboard again in suspension in the outflowing water. Rough gravel and stones are often raised by sand-pump dredgers when working on sandy shoals interspersed with larger materials; but they wear the pipe rapidly and are liable to damage or stop the pump, so that sand-pump dredgers are not commonly set to work to raise such materials. The ordinary sand pump is not suited for working in compact materials; but when fitted with revolving cutters, breaking up clay or other stiff soils at the bottom close to the mouth of the pipe, the dredger is capable of raising the

[1] The sand-pump dredger is from a drawing supplied to me by Messrs. J. and H. Gwynne of London.

loosened material [1]. The same object has also been attained by a sort of plough, fixed to the end of the pipe, which digs into the material to be lifted. When the suction pipe is fitted with a telescopic joint, or a flexible end, the sand-pump dredger can work in exposed situations, when the waves do not exceed two or three feet in height.

The earliest sand-pump hopper dredger appears to have been constructed for removing deposits of sand and silt from the port of St. Nazaire, and commenced working in 1859. It consisted of a screw steamer provided with two hoppers in its hold, and having a suction pipe hanging down on each side from amidships, and slanting down towards the stern, the two pipes being connected at the bottom by a perforated pipe resting horizontally in the bed of silt, through which the sand or silt was drawn up by the pumps [2]. As soon as the hoppers were filled, which in the two larger dredgers subsequently built had a capacity of 360 cubic yards, the engine which had worked the pumps was put into gear for working the screw, to convey the vessel to the place of deposit nearly a mile off. Each of the larger dredgers cost £6,100; and the three sand-pump dredgers raised and deposited the material at a total cost of only 2·13*d*. per cubic yard, or about 2·34*d*. per ton of this light material; whereas with a bucket-ladder dredger used for removing the more consolidated deposits close to the quay walls, the cost reached 5·57*d*. per cubic yard, or about 5·65*d*. per ton. The hopper was filled in 3½ hours : and the depositing occupied 1⅜ hours, including delays in passing through the locks. The depth of dredging at St. Nazaire ranges from 10 to 30 feet; but these vessels could dredge to over 65 feet in depth with the pipes hanging down vertically. It was found that if the perforated pipe was allowed to sink more than 18 inches below the surface

[1] Minutes of Proceedings Institution C. E., vol. civ, p. 191.
[2] 'Le dévasement du port de Saint-Nazaire,' M. Leferme, Annales des Ponts et Chaussées, 1869 (2), p. 15.

of the sand, it became difficult to haul the vessel forward, and a crust forming over the hollow produced, as much sand as water was pumped up. This early example of dredging by suction is interesting as an instance of remarkably economical work, and also as being probably the first attempt at combining a dredger and hopper in one vessel, which does not appear to have been extended to bucket-ladder dredgers till thirteen years later.

The more recent types of sand-pump dredgers have a single suction pipe in front, or at one side of the vessel, with its end dipping into the layer to be raised. Twelve sand-pump dredgers employed upon the Amsterdam Ship-Canal works, had a suction pipe, 18 inches in diameter, supported by a wrought-iron framework, and projecting in front of the bow of the vessel, which could be raised or lowered by a chain attached near the end of the pipe, and suspended from a projecting beam [1]. Each of these dredgers could raise about 1,300 tons of sand in a day, by means of a centrifugal pump having a fan 4 feet in diameter ; and the best results were obtained when the mouth of the pipe was kept 3 or 4 feet below the surface of the sand, into which it tended to bury itself too deeply. The total cost of pumping up the sand in the entrance harbour, and conveying the material 2 miles out to sea in hopper barges, was 7·23*d.* per cubic yard, the cost of each sand-pump dredger being £5,000. The same work accomplished by a bucket-ladder dredger, delivering into steam hopper barges, cost altogether 8·33*d.* per cubic yard, the original cost of this dredger having been £25,000.

A smaller sand-pump dredger, costing only £2,000, and having a fan 2 feet in diameter, and a 1-foot suction pipe, raised on the average 200 tons of gravel, sand, and stones in Lowestoft Harbour per hour, which were transported 2 miles out

---

[1] Minutes of Proceedings Institution C. E., vol. lxii, p. 23, and plate 3, figs. 7, 8, and 9.

to sea, at a cost of only 2*d.* per ton[1], which, allowing for interest and depreciation, would not amount to 3½*d.* per cubic yard raised and deposited.

An illustration is given on Plate 2, Fig. 8, of one of the sand-pump dredgers employed in deepening the channel across the bar of the Mersey in Liverpool Bay (Plate 7, Figs. 11 and 12). A centrifugal pump and suction pipe have been fitted to a 500-ton steam hopper barge; and the vessel, under favourable conditions, can raise six or seven loads in a day and deposit them 2 miles away from the bar, the filling of the hopper sometimes occupying only half an hour. Two of these dredgers removed 750,000 tons of sand from the bar in the year ending June 30, 1892, though during three months of this period only one of the vessels was at work. A much larger dredger of the same type has been constructed for the Liverpool Dock Board, with a hopper having a capacity of 3,000 tons, which it is able to fill in three quarters of an hour, in order that the deepening of the bar channel may be pushed forward more rapidly[2].

Sand-pump dredgers have been most advantageously employed in deepening the approach channels to some North Sea and Channel ports, across the sandy foreshore in front of their entrances, more especially at Dunkirk, Calais, and Ostend. At Dunkirk, the work was at first let by contract at 1*s.* 9*d.* per cubic yard raised and deposited 2⅛ miles off, and the price was eventually reduced to 7·28*d.* ; but as this price still threw an excessive burden on the port, the engineer took the work in hand with three dredgers, each costing £5,600, provided with a pump 5 feet in diameter, and having a hopper capacity of 315 cubic yards[3], whereby the total cost was reduced to about 2·1*d.* per cubic yard allowing for interest and depreciation.

---

[1] Minutes of Proceedings Institution C. E., vol. lxii, p. 45.

[2] ' The Sand Dredger Brancker,' A. Blechynden, International Maritime Congress, London, 1893, Section III, p. 52.

[3] Minutes of Proceedings Institution C. E., vol. lxxxix, p. 80.

Sand pumps possess the advantage of reducing the lift of the material to a minimum, as the outlet of the discharge pipe has only to be raised just high enough to deliver into the hopper. On the other hand, they have to raise a large proportion of water with the material, though the actual lift of this water is only the height of the discharge orifice above the water-level. Nevertheless, in spite of this disadvantage, they are able to raise sand very economically in exposed situations.

Eroding Machines. Various contrivances have been tried, having for their object the erosion, or stirring up of sand or silt, so that it may be carried away in suspension by the current.

Rakes or harrows have been dragged over shoals, so as to bring the material forming them under the influence of the current. This system, when applied to the bars at the mouth of the Danube, and of the Mississippi, failed to produce any marked increase in depth, owing doubtless to the enfeebled current being already fully charged with sand and silt, which it gradually deposits as its velocity is reduced on entering the sea. A rake, however, was used with success for cleansing the river Stour early in the nineteenth century[1]; and a similar device, dragged by a steamer over shoals of shingle in the Danube, in 1879, where the velocity of the current reaches 5 miles an hour, enabled the river to carry down the loosened shingle, and thus lowered the shoals about $3\frac{1}{2}$ feet[2]. Scrapers fixed to a frame, and towed by a steamer, have also been used for increasing the depth of the Upper Mississippi, the Missouri, and other rivers of North America: and revolving cylinders armed with spikes, drawn by horses. or by a barge hauling on a fixed rope, are used for removing weeds and stirring up the mud in the rivers and drains of the Fen country.

Screw propellers have been employed, both in Europe and America, for stirring up sandy shoals, the particles of which are carried down in suspension by the current.

---

[1] Transactions Institution C. E., vol. ii, p. 181, and plate 15.
[2] Minutes of Proceedings Institution C. E., vol. lx, p. 387, and plate 17.

A machine, called an 'eroder,' has been used in the river Witham for disintegrating and churning up compact material, so that it can be carried down, in a fine state of division, by a feeble current having a velocity of only 1½ miles an hour. The eroder consists of a circular cutter revolving very rapidly at the end of a vertical shaft, which is lowered to the bottom of the river from a steam barge carrying the engine which works the eroder[1]. It resembles in principle the disintegrating apparatus of the von Schmidt sand-pump dredger.

Jets of water, or of air under pressure, have been directed against muddy deposits and sandy shoals, driving the material upwards into the current which carries it along in suspension.

All these contrivances possess the advantage of leaving the actual removal of the material, which has been loosened or put in suspension, to be effected by the agency of the current. Accordingly, the system is economical for obtaining small increases in depth when the current is powerful, or the material very light; but it is not adapted for removing dense material when the current is feeble, nor suited for a river heavily charged with detritus. Moreover, unless the river has a surplus of depth lower down, the material transported is liable to form a shoal elsewhere in settling when the velocity of the current diminishes. The eroding principle is, therefore, only applicable under special local conditions, and where the available funds for the improvement of a river do not admit of more systematic measures, or where the requirements of navigation are very limited.

Remarks on Dredgers. The bag and spoon are still serviceable where the depth is slight, where the amount of work to be done is very small and scattered about, and where funds do not admit of a large expenditure on plant.

The bucket-ladder dredger is best suited for carrying out extensive and continuous dredging operations, such as have

---

[1] 'The Transporting Power of Water, with a Description of the Eroder,' W. H. Wheeler.

been conducted for a number of years on the Tyne, the Clyde, the Tees, and many other rivers. It can dredge materials of very varied compactness, ranging from silt to boulder clay and soft rock; and though it absorbs a large amount of power in working, owing to its cumbrous machinery, and the height to which it has to lift the material to deposit it into the sloping shoot, it is able to perform large and regular dredging works economically and expeditiously. The cost of such plant is necessarily large, and consequently the system is not well adapted for a small amount of work, unless the plant can be hired temporarily, which is generally a somewhat costly proceeding. The choice between a stationary dredger with hopper barges, and a hopper dredger, depends upon circumstances: the former plant is more costly, but it can work more continuously, provided the dredger is allowed to ground at low water, and is therefore best for a large quantity of work. The hopper dredger being self-contained is cheaper; it requires fewer hands; and it is convenient for a moderate amount of dredging, and for the relatively small amount of maintenance which may · be subsequently required. The bucket-ladder dredger is not able to dredge in considerable depths of water, being limited by the length of its ladder ; and it cannot work when subjected to wave motion. This form of dredger, however, is more extensively used than any other; and it has rendered inestimable services in river, canal, and harbour works.

The dipper dredger has been largely used in America; but though it can deepen a channe uniformly, it cannot work as continuously as the bucket-ladder dredger, or dredge at the same rate. Accordingly, the dipper dredger is better suited for dealing with detached shoals than for deepening a long length of river; and as it can only work with its full effect when making a cut about 4 feet deep, it is not adapted for the ordinary work of maintenance. It can dredge better on a rough irregular bottom, with abrupt variations in depth, than the bucket-ladder dredger; but, like that dredger, it cannot adjust its operations to the rise and fall of its vessel, resulting from waves or a swell; and the depth of its working is limited by the length of its handle.

The grab dredger is specially advantageous in raising rough obstructions, or pieces of blasted rock ; and it can work in any depth, and in exposed situations. Moreover, it can dredge in confined spaces, and close to quay walls; whilst its cost is comparatively moderate. Its vertical, discontinuous action unfits it for regular deepening operations in·rivers ; but it can be usefully applied to the removal of detached shoals of moderate extent. The grab can readily dredge sand, gravel, or other loose material ; but it requires to be made specially heavy, and to be furnished with claws, to penetrate compact or stiff ground, which can be more efficiently dredged by a bucket-ladder or dipper dredger.

The sand-pump dredger is capable of raising pure sand, in exposed situations, very cheaply ; and it can even lift compact material, if this material is previously disintegrated by a cutter. Where the material is favourable, the sand pump can dredge more economically than the ladder dredger; and it can deepen the entrances to ports, and the channels over bars, in places where ladder and dipper dredgers would often be unable to work. Moreover, sand-pump dredgers are generally cheaper than ladder dredgers; and sometimes a bucket-ladder dredger carries a sand pump in addition, so as to be able to dredge the various materials met with in the course of a long river, by the most expeditious and economical method, thereby increasing the capabilities of the dredger.

Removal of Rock under Water. As dredging machines are unable to penetrate hard rock, it is necessary first to shatter the rock, and then to raise the loose material by dredgers. The breaking up of the rock is generally effected by drilling holes in the reef from a barge, and then blasting with dynamite, a method which was adopted for lowering a rocky shoal in the navigation channel of the Tees estuary. The removal of two extensive reefs in New York Harbour, to a considerable depth, was accomplished by driving a net-work of galleries several feet below the surface of the reef, and then blowing up the roof and pillars of these galleries by the simultaneous firing by electricity of a great number of blasting charges, after which the débris was raised by grab dredgers[1]. These systems, however, are tedious and expensive; and the latter is only applicable to the exceptional case of a large area of rock having to be removed to a considerable depth.

The widening of the Suez Canal has necessitated the removal of a considerable quantity of rock under water, where the canal passes through rocky strata between the Bitter Lakes and Suez, which has led to the designing of an apparatus for shattering the rock without using explosives[2]. This apparatus consists of a long heavy iron cutter, with a steel chisel-edge at the bottom, which is let fall from a height of 5 to 20 feet on to the rock, and is raised again by hydraulic power for a repetition of the blow (Plate 2, Fig. 10)[3]. The machine used on the Suez Canal consists of a bucket-ladder dredger, with staging erected near the bow, from which five rock-breaking rams, or cutters, each weighing 4 tons, are worked; and these rams shatter the rock by a succession of blows, and the débris is then raised by the buckets. Sometimes these rock cutters, when placed singly, or two or three on a barge, are made 8 tons in weight; and they can give about thirty-five blows per hour, breaking up about 2 cubic feet of rock per blow. This system is also being used for the removal of a portion of the rocky barrier across the Danube, known as the 'Iron Gates.'

[1] 'Blasting Operations at Hell Gate, New York,' L. F. Vernon-Harcourt. Minutes of Proceedings Institution C. E., 1886, vol. lxxxv, p. 264.

[2] 'The Removal of Rock under Water without Explosives,' F. Lobnitz. Minutes of Proceedings Institution C. E., 1889, vol. xcvii, p. 369.

[3] The illustration of three rock-breaking rams on a barge was taken from some drawings furnished me by Messrs. Lobnitz & Co. of Renfrew.

The remainder of the obstruction forming the ' Iron Gates '
is being perforated by rock drills, worked by steam, inside
iron tubes resting on the rock.   When the hole has been bored
to its full depth, it is cleared out by a jet of water, and
a dynamite cartridge is inserted, and a series of charges
are exploded by an electrical battery.   During the boring
of the holes, the barge carrying the apparatus is raised slightly
out of water, resting upon four props standing on the bed of
the river, so as to be secure from wave motion [1].   This boring
apparatus has for many years given good results in the
removal of rocks in the St. Lawrence.

DISCHARGE OF DREDGED MATERIALS.

The actual cost of dredging depends upon two distinct
operations ; first, the cost of dredging and raising the material
to the surface ; and, second, the cost of conveying it to its
destination.   It has been already indicated that, in certain
circumstances, economy or convenience may be attained by
accomplishing both operations with the same vessel, by using
a hopper dredger.   Generally the dredged material, having
been discharged into barges alongside the dredger, is conveyed
out to deep water and deposited through a hopper ; or, if
the site is at a distance from the sea, it has to be deposited
at a suitable place along the banks, by manual labour.

Discharging by a Chain of Buckets.   The cost of unloading
barges by manual labour is necessarily large ; and if the
material is very soft. the operation is difficult.   Accordingly,
on the Danube regulation works, at Vienna, the material
was removed from the barges by a chain of buckets, similar
to those employed in bucket-ladder dredgers, supported on
a staging on the bank, and was discharged into wagons running

[1] 'La Régularisation des Portes de Fer,' Béla de Gonda, Congrès international
de Navigation intérieure, Paris, 1892.

on to the staging [1]. The same system was also adopted at the enlargement works of the Ghent-Terneuzen Canal. At the Panama Canal works, a floating bucket-ladder dredger raised the dredged material out of the barges and discharged it into a long shoot; and similar plant is at work on the Maas.

Discharging direct into Wagons. Another plan of reducing the cost of transport of dredged materials, adopted a few years ago in France, consists in discharging the material, brought up by the buckets, direct into wagons placed on the barges, instead of into the barges themselves [2]. Having erected a suitable landing stage, the wagons on reaching the shore can be at once run off on to a railway, thus dispensing with the double shifting of the material necessitated by the ordinary processes. A saving of 2*d.* per cubic yard was effected by this system.

Discharging through a Long Shoot. Where the dredged material has only to be conveyed a short distance to the side banks, special contrivances have been adopted with great advantage. Thus on the Suez Canal works, long shoots were employed for conveying the dredged material, as it fell from the buckets, away to the banks. The shoots were made as much as 230 feet in length (Plate 2, Fig. 7). They were semi-elliptical in section, 5 feet wide, and were hinged to the dredger to admit of an alteration in their inclination, which was accomplished by raising or lowering the supports on the lighter between the dredger and the bank. Mud was carried away with hardly any inclination of the shoot; sand mixed with water was conveyed away with an inclination of 1 in 20; but in dealing with sand and shells, or clay, a pair of endless chains were employed, together with a stream of water, for dragging the material down the shoot.

Similar long shoots were employed on the Panama Canal works where the banks were low enough. These steel shoots were 131 feet long, with an inclination of 1 in 20, and were

---

[1] Annales des Ponts et Chaussées, 1875 (1), p. 405.   [2] Ibid. 1880 (1), p. 29.

supported by steel wire cables passing over sheer-legs erected on a vessel fastened alongside the dredger on the side of the shoot, and counterpoised, on the other side, by a vessel also fastened to the dredger and loaded with ballast[1]. The material was carried along the shoot by a stream of water, and was discharged 11 feet above the water-level. Long shoots have also been made to extend 200 feet out on each side of the dredger, supported by wire cables from sheer-legs erected on each side of a wide barge carrying the dredger.

Discharging by a Pump and Floating Tubes. The material dredged from the sandy bottom of Lake Y and Wijker Meer, in constructing the Amsterdam Ship-Canal, was discharged on to the side banks by mixing it with water, and forcing it, in a semi-fluid state, through long wooden tubes floating on the surface of the water. A Woodford pump, $3\frac{1}{2}$ feet in diameter and performing 230 revolutions per minute, was fixed to the side of the dredger at the level of the water, having a reservoir at the top in which the dredged material falling out of the buckets was received, and an inlet at the bottom for admitting the water. The material, mixed with water, was forced by the pump through the tubes, which were 15 inches in diameter, and were buoyed up by floats on the surface. The line of tubes, connected together by leathern joints, extended in some cases to a distance of 300 yards; and the material could be delivered 8 feet above the water-level[2]. A large dredger was thus enabled to excavate, and deposit on the banks, about 1,300 cubic yards per day of twelve hours, at a cost of about 2d. per cubic yard.

A similar contrivance was adopted at the Ghent-Terneuzen Canal enlargement works; but in this case a centrifugal pump was joined on at about the middle of the floating tube, which increased the discharge, and raised the material, consisting of sand and silt, to a height of 28 feet, whence it

---

[1] 'Le Génie Civil,' vol. v, p. 17.
[2] Minutes of Proceedings Institution C. E., vol. lxii, p. 6.

was led into a shoot which conveyed it considerably further[1]. By the addition of a second pump to the tube, the material was eventually discharged about five furlongs away from the dredger.

At the same canal works, a combination of the pump with a long shoot was also used for conveying the dredgings a distance of about 180 feet, and discharging them at a height of 13 feet above the water-level. The material was forced through the supported tube, by a pump discharging a rapid stream of water amounting to about three times the volume of the material.

EXCAVATORS.

The main portion of the excavations for new river channels, canals, basins, and docks, has generally to be executed before the water can be admitted ; and dredging is only available for forming the entrance channels and the waterway through lakes, and for the final deepening and maintenance. Accordingly, machines have been designed for effecting large excavations on land, on the same principles that dredgers accomplish them under water. Some excavators were designed for the Suez Canal works ; and though the Panama Canal works have little to recommend them in their present stage, they undoubtedly gave a stimulus to improvements in excavators, from which subsequent works, like the Manchester Ship-Canal, have largely profited.

Bucket-Ladder Excavator. A machine was employed on the Suez Canal works, which carried an excavating apparatus like the chain of buckets of a dredger, and moved on wheels like a locomotive, being propelled by one steam-engine, whilst another worked the buckets. Seven of these excavators removed about eight million cubic yards on the Suez Canal ;

---

[1] 'Le Canal de Terneuzen-Gand, et ses Installations Maritimes.' O. Bruneel and E. Braum.

and similar machines were subsequently used on the Danube regulation works, the enlargement of the Ghent-Terneuzen Canal, and the Tancarville Canal. These excavators, supported on four axles with three wheels each, ran on three rails laid on the surface of the ground near the edge of the top of the cutting, and excavated slices off the slope, like a plane, with their ladder projecting down the cutting, and inclined at the angle of the slope. The buckets raised the material and deposited it in wagons running on a line of rails alongside the excavators; and one of these excavators could remove about 3,000 cubic yards from the sandy cutting of the Suez Canal, on the average, in a day. Several excavators, similar in type, though differing in details, were employed on the Panama Canal works. Seven bucket-ladder excavators, of French and German make, have been used on the Manchester Ship-Canal, each costing about £2,400, and weighing 70 to 80 tons[1]. Their buckets cut the slope inwards towards the machine; and being merely curved round at the bottom, and open towards the chains, they discharge their contents downwards into the wagon standing below in turning over the top tumbler. Each of these excavators can remove 1,400 to 1,500 cubic yards per day on the average, the maximum amount raised in a day being 2,500 cubic yards. The excavation with these machines has cost about 1½*d.* per cubic yard, exclusive of interest and depreciation. These excavators, however, whilst working very economically in light soft soils, have proved unable to deal with heavy or stiff materials, such as soft rock, stiff clays, boulders, or embedded trunks of trees. Owing to their great weight, heavy steel rails and stout sleepers are required to carry them; and the road needs frequent repairs, especially on soft ground in which some of the machines are in danger of settling and tipping over into the cutting, an accident which

---

[1] 'Mechanical Appliances employed in the Construction of the Manchester Ship-Canal,' E. Leader Williams. Minutes of Proceedings Institution M. E., July, 1891.

occurred once or twice on the canal works. Moreover, they are expensive to remove from one cutting to another. They possess, however, the advantage of lifting the material out of the cutting, and thus reduce the cost of haulage.

Steam Navvy. The type of excavator which works a large cylindrical bucket, with movable bottom, at the end of a long beam, like a dipper dredger, is known generally as a steam navvy, and is extensively employed in large canal and dock excavations (Plate 2, Fig. 9)[1]. Sometimes the bucket is worked from a strongly framed revolving jib, as shown in the illustration; and in smaller types, the bucket is connected to the jib of a revolving steam crane. The advantages of the large steam navvy are its considerable capacity for work, and its power of excavating all kinds of material except hard rock; and the smaller machine has the merits of comparative lightness, so that it can run on the ordinary temporary roads, smaller cost, and the availability of the steam crane for other work when not required for excavating.

A large steam navvy of American manufacture, running on a bogie frame with eight wheels, and armed with a steel bucket of 2⅛ cubic yards capacity, was used in the deep Culebra cutting of the Panama Canal, and excavated 520 cubic yards per day under difficult conditions, its capabilities being stated to be five times as great[2]. An English steam navvy also employed at the Panama Canal, weighing 32 tons and running on two pairs of wheels, with a bucket holding 1⅛ cubic yards, excavated nearly 800 cubic yards per day on the average[3]. Altogether, not less than seventy-two excavators of different kinds were at work on the Panama Canal shortly before the stoppage of the works, thirty-nine of which were excavating in the Culebra section only a little over two miles in length.

[1] The illustration of a Steam Navvy has been reduced from a drawing furnished me by Messrs. Ruston, Proctor, and Co. of Lincoln.

[2] 'Le Génie Civil,' vol. v, 1884, p. 393.   [3] Ibid. vol. xiii, 1888, p. 131.

H

About one hundred excavators were in use at the same time on the Manchester Ship-Canal, fifty-eight of which were steam navvies of the larger type, costing about £1,200 each. The buckets of these steam navvies have a capacity of from $1\frac{1}{4}$ cubic yards for boulder clay, up to $2\frac{1}{4}$ cubic yards for sand; and the average output of one of these excavators is 600 to 700 cubic yards per day of ten hours, though a machine with the largest bucket has excavated 2,000 cubic yards in a day when working in sand. These navvies can excavate a trench about 25 feet deep, and from 50 to 60 feet wide at the top. They travel on a gauge of $10\frac{1}{2}$ feet; their weight of 32 tons necessitates a very firm road; and they tend to settle on soft ground. The steam-crane navvy is readily moved, it can turn right round and excavate in any direction, its average output is from 400 to 600 cubic yards per day, and its cost is £800 to £1,050. The excavation with steam navvies on the canal cost about $2d.$ per cubic yard.

Grab Excavator. As the grab dredger is commonly worked by a crane placed on a barge, it is equally adapted for working on land from a movable steam crane. The grab excavator is advantageously used for raising dredged material from barges, and depositing it on land at the back of a quay wall, and also for excavating in cramped spaces; but its inability to penetrate easily hard or stiff material renders its application limited compared with the other forms of excavators. These machines were used on the Manchester Ship-Canal, mainly for excavating foundations, and for removing materials under water; and in suitable soils, they have been able to raise about 300 cubic yards per day. The grab bucket is a comparatively cheap machine; and it can be readily worked by a steam crane, which is available at any time for other purposes, and can also be used for loading or unloading vessels when its work of excavating is terminated.

Remarks on Excavators. The steam navvy is specially adapted for opening out the advanced trench of a cutting from the bottom; whilst the bucket-ladder excavator serves best for enlarging the cutting from the top. Some bucket-ladder excavators, however, were made for the Panama Canal works, in which the ladder of buckets, placed nearly horizontally, projected in front of the truck which carried it at the bottom of the cutting, so that these machines could open out the trench of a cutting, 30 feet wide, like a steam navvy; but though possessing the advantage of working in a continuous manner, they were less stable than a steam navvy when the ladder was turned round towards the side of the cutting. The grab excavator is only suited for special work in loose soil in limited spaces; the bucket-ladder excavator has a considerable capacity for work in moderately soft or loose strata : whilst the steam navvy can excavate any material which is not compact enough to require blasting.

Steam navvies and bucket-ladder excavators are indispensable for expediting the work of deep extensive cuttings, such as are involved in the construction of ship-canals; and they are especially valuable where manual labour is scarce or costly, and more particularly where an unhealthy climate entails peculiar risks to human life, as was encountered at the isthmus of Panama.

CHAPTER V.

LOCKS; AND FIXED, AND DRAW-DOOR WEIRS.

Locks:—Description of Parts; Construction, varieties; Lock-Gates, single and double, wooden and iron, strains, support, movement. *Weirs:*—object, classification. *Fixed Weirs:*—Description, disadvantage; Oblique Weirs, object; Angular and Horse-shoe Weirs, advantages, forms; Construction; Anicuts, methods of construction, instances, defects. *Draw-door Weirs:*—Description, object; Teddington Weir; Indian Dams; Draw-doors sliding on Free Rollers, instances, effects, for regulating tidal flow on Thames and Weaver; Segmental Gate Weir, description, object, instances. Concluding Remarks.

THE object and advantages of the canalization of rivers have been explained in Chapter III; and it is proposed in the present chapter to describe some of the works by which this canalization is carried out, which consist essentially of weirs which keep up the water-level of a river in successive reaches, and locks which enable vessels to pass from the water-level of one reach to the water-level of the adjoining reach.

LOCKS.

A lock contains a chamber, placed at the junction of two reaches, in which the water can be raised or lowered so as to be on a level with either the upper or lower reach. The lock-chamber is usually closed by a pair of gates at each end; it is filled by letting in water from the upper reach through sluices in the upper gates or side walls, and emptied by letting it out through similar sluices at the lower end. A vessel, admitted into the lock when the level of the water in the lock is the same as in the reach which the vessel

has just traversed, can be thus easily raised or lowered to the
level of the adjoining reach.

Description of a Lock. The parts of a lock are shown
on Fig. 11. They consist of two side walls, AA, with
curved ends to facilitate the entrance and exit of boats; and
two pairs of gates, BB, at each end of the lock-chamber,
C, shutting at the bottom against a sill, DD. The bottom
of the lock-chamber is covered by the invert, E, which, as
its name implies, is generally made in the form of an inverted
arch to support the side walls, exposed as they are to
constant variations in pressure, and to secure the chamber

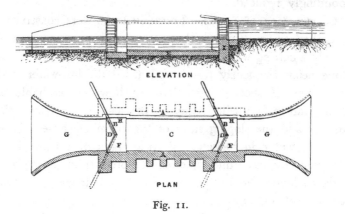

Fig. 11.

against injury or leakage from below. The portion of the
floor of the lock, FF, over which the gates move is called
the gate-floor. Beyond the gate-floor, at each end of the
lock, is the apron, GG, which is generally protected from
scour by pitching or masonry. The vertical post on which
each gate turns, called the heel-post, stands upon a stone
called the heel-post stone, and fits into a vertical curved
groove in the lock wall, formed by courses of carefully
dressed stones, which from their shape are called hollow
quoins (Fig. 12, page 105). A recess, HH, is formed in
the lock wall, on each side of the gate-floor, into which

the gate is drawn, when opened, so that it may be out of the way of boats passing through the lock; it is called the gate recess, and is terminated at one end by the hollow quoins, and at the other end by another set of dressed stones called square quoins.

In locks leading to docks, or on rivers where the fall of the bed is not abrupt, the upper and lower sills are placed at the same level; but on canals, and on rivers where the original bed is altered, the upper sill is placed on the top of the lift wall, *I*, at the level of the bottom of the upper pool, and the depth of the upper pair of gates is correspondingly reduced.

Construction of Locks. Formerly locks were constructed of timber, or sometimes with masonry walls at each end, and merely a cutting with sloping sides for the lock-chamber, the slope being frequently terminated about the low-water level by a line of sheet-piling. Timber, however, is liable to decay, and to be washed away by the rush of water from the sluices; and the sloping sides of the chamber cause a great waste of water in locking. Vertical walls of brickwork or masonry are, accordingly, often adopted throughout; and in large locks, the invert, gate-floors, and aprons, are constructed of masonry, and the sills and hollow quoins of granite. Locks are made just large enough to admit the largest class of vessels employed on the special navigation; but where the size of the vessels varies considerably, an intermediate pair of gates is sometimes introduced, so that a shorter lock-chamber may be provided for small vessels, thus saving both time and water in locking. Suresnes lock, the first lock on the Lower Seine below Paris, furnishes an instance of this arrangement, the narrower of the two locks being provided with an additional pair of gates, dividing the lock into two portions, 133 feet, and 331 feet long respectively. The large traffic, however, on the Lower Seine has allowed the same object to be accomplished at other

places by providing locks of different sizes, as for instance
at Poses, where the three locks are of different lengths and
widths (Plate 4, Fig. 4). In the locks at both ends of the
Amsterdam Ship-Canal, both methods are resorted to for
accommodating the variety of traffic (Plate 12, Figs. 3 and 4).
The five locks on the canalized Main, below Frankfort,
have been recently greatly lengthened by adding a third
pair of gates in the cut below the original lock, providing
a second lock, 820 feet long, capable of admitting six of
the largest Rhine boats two abreast, with their tug occupying
the original lock. These smaller locks, 279 feet long and
34½ feet wide, will still serve to accommodate the steamers
of 1,300 tons navigating the Rhine, for which they were
designed (Plate 4, Fig. 5). When a lock is intended to hold
a double row of large vessels, the entrance is only made
wide enough to admit the broadest vessel, and the lock-
chamber is widened out.

LOCK-GATES.

The smallest locks are closed by a single gate at each
end. There is, however, an advantage in closing the lower
end of the lock with a pair of gates, as a single gate requires
more space for opening, and thus either reduces the available
length, or necessitates an increased length of lock, and a corresponding increased expenditure of water in locking. Except
in small locks, both ends of the lock-chamber are closed
by a pair of gates, each gate being rather longer than half
the width of the passage, and meeting at an angle in the
centre line of the lock. Each gate turns on a heel-post,
and has a meeting- or mitre-post fixed at its other extremity,
which, when the gates are closed, shuts against the mitre-
post of the opposite gate. The mitre-posts of the two gates
are formed so that their meeting surfaces may be in close
contact when the gates are closed; and a horizontal sill-
piece at the same time shuts closely against the sill.

Wooden Lock-Gates. Lock-gates across small locks, and in a moderate depth of water, are generally made of wood; and at the Amsterdam Canal, wooden gates have been adopted for the inner gates of the locks having a width of 60 feet; but in this instance the variations in the level of the water are slight, and consequently the pressure is moderate. English oak is the best material for wooden gates; but where the gates are liable to be attacked by the teredo navalis, greenheart timber must be used, though it is costly to work and liable to split. The Manchester Ship-Canal lock-gates have been constructed of greenheart.

Wooden gates are formed of a series of horizontal beams, connecting the heel-post and mitre-post; against which vertical or diagonal planking is fastened, and, being made watertight, acts as the skin of the gate. The beams are placed closer together towards the bottom of the gate, in order to sustain the increased pressure due to the greater depth of water.

Iron Lock-Gates. Iron gates have generally a double skin of wrought-iron plates, riveted to horizontal ribs or girders; and they are strengthened in addition by vertical ribs. The skin plates are increased in thickness, and the horizontal ribs are placed closer together and strengthened towards the bottom of the gate, to allow for the increased strain. The lock-gates of the Amsterdam Ship-Canal pointing towards the sea, have been constructed of iron, to afford additional security against any inroad of the sea.

Strains on Lock-Gates. The strains on a single straight gate shutting across a lock result merely from the pressure of the head of water against its up-stream face, which increases in proportion to the depth. The horizontal beams of the gate act like girders, supporting an evenly distributed load which imparts a strain equal to half the whole pressure exerted at the centre.

When, however, two straight gates meet at an angle in the

centre line of the lock, each gate, besides being subjected to the water pressure on its inner face acting as a transverse strain, has also to bear a compressive strain in the direction of its length, due to the pressure of the opposite gate against its mitre-post. This latter strain is equal to half the water pressure on the opposite gate, multiplied by the tangent of half the angle between the lines of the two gates when closed. This angle depends upon the proportion which the projection of the sill, AB, bears to the span, CD (Fig. 12), which is termed the rise of the gates. The

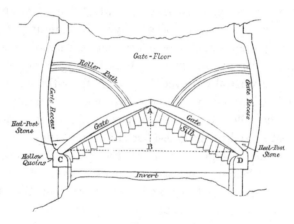

Fig. 12.

accompanying sketch shows a pair of gates having a rise of one fourth; and the tangent of half the angle CAD, is $CB \div AB = 2$, showing that the rise of a pair of gates exercises an important influence on the compressive strain transmitted by one gate to the other, which varies inversely with the rise. Thus in the old Sparndam lock-gates, with a rise of only one sixteenth of the span, the compressive strain amounted to four times the water pressure on one gate. On the other hand, a large rise increases the length of the gates and reduces the available length of the lock.

Support of Lock-Gates. The weight of a lock-gate is supported by the pivot on which the heel-post turns, and by the anchor strap which encircles the top of the heel-post. The pivot is fastened to the heel-post stone; and a socket at the bottom of the heel-post fits over the pivot. The anchor strap is fastened to the anchor, formed of a cast-iron frame, in the form of a sextant, bedded into the masonry and further secured by tie bolts. A roller also is generally placed at the bottom of large gates to assist in supporting the weight of the gate. Its axis is placed in the vertical plane passing through the axis of the heel-post and the centre of gravity of the gate; and the proportion it bears of the weight depends upon its distance from the centre of gravity. In straight gates, the roller is placed under the gate; but in curved gates it is partially clear of the gate, and its bearing is adjusted by means of a vertical rod over it, at the back of the gate, which is raised or lowered by a screw or wedges near the top of the gate. The roller, which is slightly conical, runs on a cast-iron roller-path fixed to the gate-floor (Fig. 12, p. 105).

Wherever a large portion of the gate remains always immersed when it is opened and closed, as in the case of the upper gates of locks, and the Amsterdam Ship-Canal lock-gates, a roller may be dispensed with even for large gates, owing to the buoyancy of the immersed gate, and the small weight, consequently, which has to be supported by the pivot.

Methods of moving Lock-Gates. The smaller wooden lock-gates are opened and closed by pushing against the balance beam, which has a heavy butt end projecting over the side wall of the lock (Fig. 11, p. 101), and is fastened to the top of the heel-post and mitre-post, and serves to balance the gate on the pivot. The larger gates are moved by chains fastened to the gate, and worked by capstans or hydraulic machinery. The employment of rollers renders it advisable to attach the chains some way down on each side of the gate, in order to facilitate the motion and to avoid straining the gate. This necessitates the construction of chain passages in the side walls for the opening and closing chains, with rollers for directing and guiding them. When the gates are constructed without rollers, the chains can be attached near the top of the gate, and worked from the top of the side walls, as at the Amsterdam Canal locks. A still better plan in such cases, where there is ample space at the sides, is to fasten a long, flat, toothed iron bar to the top of each gate, which opens or closes the gate on being moved by a pinion on the top of the side wall.

Further particulars relating to lock-gates, with illustrations, will be found under the heading 'Dock-Gates[1]' in the part of 'Harbours and Docks' descriptive of docks, to which the subject of large lock-gates more strictly belongs.

WEIRS.

Weirs constitute the barriers which, placed at suitable intervals in the bed of a river, raise the water-level of the river, converting it into a series of steps of nearly level reaches, the differences in level of which are surmounted by means of locks which form the necessary adjuncts to weirs in a continuous river navigation. The discharge of the river passes over, or through the weirs; whilst the vessels pass in ordinary times through the locks. The weir stretches across the main channel of the river; and the lock is generally placed in a minor channel separated off by an island, or in an artificial side cut, so that the vessels may be kept clear from the rapid current caused by the discharge of the river falling over the weir. The former plan has been adopted on the Lower Seine (Plate 4, Fig. 4); and the latter method has been resorted to on the River Main (Plate 4, Fig. 5). On the

[1] 'Harbours and Docks.' L. F. Vernon-Harcourt, pp. 443–453, and Plates 15 and 16.

Upper Seine, on the contrary, in the absence of islands, the
locks have been placed close to one of the banks, alongside
the weir; but the security of the vessels has been provided
for in these cases, by removing the upper end of the lock
as far as practicable from the weir by putting the lower end
in a line with the weir (Plate 4, Fig. 3).

Weirs may be conveniently divided into three classes;
namely, 1. Fixed, or Solid Weirs; 2. Draw-door Weirs; and
3. Movable Weirs. No portion of the first type of weir is
removable; and it forms a solid, permanent obstruction in the
channel of a river. Draw-door weirs, whilst affording an
equally efficient barrier as a solid weir during the low stage
of a river, can be more or less opened in flood-time to facilitate
the discharge of the river. Movable weirs serve equally as
well as solid and draw-door weirs to retain the water-level
of a river during the dry season; and they possess the
advantage, not only of being able to regulate the flow of
a river, and to facilitate the discharge of flood waters like
draw-door weirs, but also of leaving the river entirely open
in flood-time.

FIXED WEIRS.

A fixed weir is a solid wall, or embankment, placed across
a river, which affords no passage for the discharge of the
stream till the water has risen above it. This kind of weir
is very efficient in preserving a fixed minimum water-level
above it; but it is incapable of regulating the flow, and
is not suitable for the rapid discharge of flood waters. As
the whole drainage of the district above, except the water expended in locking, must pass over the solid weir, if no other
outlet is provided, such a weir, raised to the level required
for maintaining the proper navigable depth in dry summer
weather, diminishes considerably the waterway which existed
for the discharge of the river previous to its erection.

Endeavours have been made to mitigate this disadvantage

of fixed weirs. Sometimes the sill of the weir is placed below the summer water-level, and is raised in time of drought by placing planks along the top of the sill, which are kept in place by upright stakes. Another method of facilitating the discharge consists in making the crest of the weir longer than the ·average width of the channel, by placing the weir in a wide part of the river, or giving it a slanting direction across the river.

Oblique Weirs. Some weirs erected across the Severn, in 1842 [1] (Plate 3, Figs. 7 and 8), were placed obliquely to the stream, in wide places on the river, so that the length of the weirs, and consequently the capacity for discharge over them, was increased. The Severn has high banks and a rapid fall in the part where the weirs were introduced, so that solid weirs are less objectionable there than they would be in rivers having low banks and a small fall. The river, under these conditions, is able to rise a considerable height above the top of the weir without overflowing its banks ; and when it does so, the obstruction to the discharge presented by the weir is proportionately less. The length of the weirs on the Severn was made such that the section of the waterway over the weir in flood-time was equal to the sectional area of the river above; but, owing to the less favourable form of the shallower section over the weir, and the impediment occasioned to the flow by the change in form of the channel, and in the direction of the current at the weir, the discharge must always be somewhat checked at that part. Also, when the velocity of approach of the current increases, as it does in flood-time, the advantage gained in length by the oblique position of the weir is somewhat neutralized ; for the filaments of the water tend to flow over the sill of the weir in a direction approximating to the axis of the river, instead of at right angles to the weir.

Simultaneously with the introduction of these weirs on

[1] Minutes of Proceedings Institution C. E., vol. v, p. 340.

the Severn, the channel of the river was greatly improved, and numerous rocky fords were removed ; so that in spite of the obstructions presented by the weirs, the discharge of the river in flood-time was on the whole facilitated ; whilst a better navigable depth in dry weather was secured by the deepening of the river, and by the weirs.

Angular and Horse-shoe Weirs. As straight weirs placed obliquely across a river tend to divert the main current, and consequently the deepest channel, towards the bank against which the up-stream end of the weir rests, other forms of obliqueness are frequently given to fixed weirs. They are sometimes constructed in two straight lines, joining in the centre of the river at an angle pointing up stream (Plate 3 Fig. 9) ; or sometimes a curved form is adopted, convex on the up-stream side, so that a sort of horse-shoe fall is formed. These forms, whilst increasing the length of the weirs, direct the main stream into a central channel ; but the clashing of the currents is liable to produce eddies, and cause undue scour below the centre of the weir, so that the bottom must be protected at this part.

The torrential River Lot in France was canalized many years ago, for a distance of 185 miles, down to its confluence with the Garonne, by over seventy fixed weirs in broad places in the river, with some of the locks adjoining the weirs, and some inside cuts [1]. Several mills had been previously established on the river, retaining its waters by dams ; and, moreover, the high banks of the river rendered inundations exceptional, so that under these conditions fixed weirs were unobjectionable. The earlier weirs erected on the River Lot were made angular (Plate 3, Fig. 9); but this form proved inconvenient for navigation, owing to the eddies produced, forming back currents towards the lock on the down-stream side. An oblique weir, inclined up-stream from the bank on one side of the river, and connected

[1] Annales des Ponts et Chaussées, 1865 (1), p. 151, and plates 103 and 104.

with the lock on the other bank by a portion of weir at right angles to the stream, directs the currents in a more satisfactory manner (Plate 3, Fig. 11).

Construction of Fixed Weirs. Fixed weirs are frequently formed of a mound of rubble stone protected on the slopes by pitching (Plate 3, Fig. 8). The up-stream slope is made steep, and the down-stream slope very flat ; and the pitching, or a layer of rubble stone, is continued beyond the foot of the down-stream slope to protect the river-bed from scour. The top of the weir is capped by a sill of timber or stone. A row of sheet-piling along the line of the sill, and sometimes also along the toe of the down-stream slope, secures the weir from being undermined by the water (Plate 3, Figs. 8 and 10). Sometimes the weir is formed by a thick masonry wall, battered on both faces, and with a slight depression towards the current on the top to facilitate the passage of the water over the sill (Plate 3, Fig. 12).

Anicuts. Large fixed weirs have been erected across rivers in India, in order to retain and divert the water for irrigation. These weirs, termed anicuts in the Madras presidency, keep back the water during the dry season; and in flood-time the river flows over the weir. The difference between the low-water level and the flood-level is so great, that the weir, whilst high enough to retain a sufficient quantity of water in the dry season, is submerged to a considerable depth during the rainy season, so that it offers less impediment to the flow in flood-time than weirs placed across rivers experiencing smaller variations in level.

In olden times the natives used to erect embankments of sand or loose stones at the close of the rainy season, which served to keep back the water when the river was low, and were washed away by the first flood. These embankments, accordingly, had to be reconstructed every year ; but as soon as these works were entrusted to English engineers, they were made more durable.

Anicuts are now generally constructed of dry rubble, concrete, or rubble stone set in cement, protected on the face
by pitching or ashlar masonry set in cement. The up-stream
side of the anicut is often built nearly vertical; and the down-
stream side has a flat slope, terminated at the base by a stone
apron laid on the river-bed to prevent scour (Plate 3, Figs.
13 and 14). The foundations are varied according to the
nature of the river-bed; but generally trenches or well-
foundations, filled with masonry or concrete, are carried
some feet below the bed, right across the river, under each
toe of the anicut, and sometimes also under the sill, to
secure the anicut from being undermined. Walls are built
on these foundations, and the intermediate spaces are filled
with loose rubble. The Okhla anicut furnishes an example
of a work in which no foundations were carried below the
bed of the river; but it is very wide in cross-section,
having been considerably enlarged beyond the original
design[1].

Some of these anicuts are very long. For instance the
Godavery anicut across the River Godavery (Plate 3, Fig. 14),
with its sill 12 feet above the river-bed, and crossing four
separate branches of the river, has a total length of nearly
2½ miles; and the Dehree anicut across the River Sone
(Plate 3, Fig. 13), 8 feet high, is 2⅛ miles long.

Silt and sand brought down by these rivers tend to accumulate above an anicut, raising the river-bed; and sometimes
the deposit eventually chokes up the entrance to the irrigation
canal, which is usually constructed a short distance above the
anicut. Moreover, an anicut, like an ordinary fixed weir,
is not suited to discharge a sudden flood, or to modify the
levels of the water above it. Sluices, accordingly, closed by
gates or draw-doors, have to be constructed in anicuts to
scour away the silt and to carry off any excess of water.

[1] 'Professional Papers on Indian Engineering,' Second series, vol. vi, p. 239,
and plate 30.

DRAW-DOOR WEIRS.

A draw-door weir consists of a row of piles, frames, or piers, placed generally at regular intervals across a portion of a river, with a series of panels, doors, or planks, to close the spaces between them. These doors or panels, sliding in vertical grooves in the piers, can be raised or lowered from a foot-bridge above, thus opening or closing the weir. Draw-door weirs, accordingly, provide a better discharge in flood-time than overfall weirs; they also furnish a means of regulating the level of the water above, and of scouring the channel near the weir.

In mill-streams, a fixed weir is usually placed across the bank where the artificial cut to the mill diverges from the old channel; and the natural fall of the stream having been reduced, and the sill of the fixed weir placed at a high level in order to keep up the water above the mill, draw-door weirs are provided to discharge the surplus water when the mill-gates are closed, or in flood-time. The doors of these weirs are generally formed of wooden panels sliding in grooves between masonry side walls and intermediate piles; and they are raised by a rack and pinion, like the sluice-gates in small lock-gates.

Similar draw-door weirs, on a larger scale, are commonly placed across a portion of a river channel where a fixed weir is situated, so as to regulate and increase the discharge. Several of these weirs with large draw-doors, sliding between piers and raised by chains passing round overhead pulleys, have been introduced in the last few years in the weirs on the Upper Thames, and have proved very valuable in reducing the height of the floods.

Teddington Weir. The weir on the Thames at Teddington, at the limit of the tidal flow, furnishes perhaps the best example in England of a large composite fixed and draw-door weir placed obliquely across the river. It is divided into four separate bays, and has a total length of 480 feet. Along the three down-stream bays, strong wrought-iron frames, formed of plates and angle-irons riveted together, are fixed at regular intervals, and carry a foot-bridge on the top (Plate 3, Figs. 15 and 16). The two side bays are fixed weirs. The two central bays, 172½ feet and 69¾ feet long respectively, are closed by large draw-doors. The longer bay has twenty-three draw-doors, 6 feet high and 7⅔ feet wide, formed of plate-iron strengthened with angle-irons; and the other bay has twelve draw-doors[1]. The doors slide in vertical grooves at the sides of the frames. A small tram-road is laid along the foot-bridge, on which a crab runs which lifts the draw-doors by means of a chain.

The summer flow of the Thames passes over the fixed weirs, which keep up the river to the requisite navigable level; and the discharge during the winter, or in flood-time, is provided for by raising the draw-doors. The additional opening thus provided does not extend down to the bed of the river, for the draw-doors with their frames, are placed on a rubble mound (Plate 3, Fig. 15); but this deficiency in depth is compensated for by the great width of the weir.

Indian Dams. In constructing weirs for irrigation across rivers in northern India, bringing down large quantities of sand or silt, it has been found expedient to erect piers in the central portion of the weir, with openings between them, about 10 feet wide, which are closed, on the approach of the dry season, by planks or balks lowered horizontally, or by gates sliding in grooves down the sides of the piers. The stone piers rest on a masonry flooring carried across the river; and the weir is built solid for a certain distance from each bank. These partially open weirs are called dams in India. In more recent dams, the openings between the piers have been made considerably larger, and are closed

---

[1] 'Fixed and Movable Weirs.' L. F. Vernon-Harcourt. Minutes of Proceedings Institution C. E., 1880, vol. lx, p. 25, and plate 1, figs. 11, 12, and 13.

by gates, turning on horizontal axes at the level of the flooring, which resemble a form of movable weir, and will therefore be described in the next chapter. These dams, whilst generally costing less than solid weirs, present less impediment, when open, to the flow of the river, and prevent the accumulation of silt and sand above the weir.

Free Rollers applied to Draw-doors. In raising a draw-door, a considerable amount of friction is generally experienced at the surfaces of contact along which the two ends of the door slide, especially when there is much pressure of water against the door. This friction has been considerably reduced by making each end of the sluice-gate slide upon a vertical row of free rollers hung on a suspended chain in a recess in each side pier; one end of the suspending chain being attached to the sluice-gate, and the other end to the side wall, causing the rollers to travel up and down at half the rate of the gate (Plate 3, Figs. 17 and 18). A watertight joint is made by a suspended rod at each end, pressed against the side wall on the up-stream side of the draw-door, so that the door itself merely rests against the rollers.

Two large draw-door weirs on this principle have been erected at Belleek and Ballinasloe in Ireland, across the rivers Erne and Suck, to control the drainage works connected with these rivers[1]; and a half-tide weir has been erected across the Thames a little below Richmond, which is closed by three draw-doors of 66 feet span working on free rollers.

The Belleek weir, erected in 1883, consists of four openings between masonry piers, closed by draw-doors, 31 feet long and 14½ feet high, and weighing 13 tons each, which can be raised 9 feet above the sill on the bed of the river by one man, against a water pressure of about 100 tons, by means

---

[1] A few particulars about these weirs, or sluices, have been given me by their designer, Mr. F. G. M. Stoney; and I have also had the opportunity of seeing them worked.

of gearing placed on a girder bridge resting on the piers. A turbine on the right bank, worked by the fall of water at the weir, can raise the four draw-doors simultaneously. This weir has enabled a rocky barrier to be removed from the river ; and, whilst holding up the water-level in Lough Erne and the adjacent Ulster Canal for navigation, this work has reduced the rise of the floods from 10 feet to 7 feet by the improved waterway provided, and by raising the draw-door on the approach of a flood.

The Ballinasloe weir was erected in 1885, across the up-stream side of an arched masonry bridge crossing the river Suck, having four spans of 25 feet closed by draw-doors sliding on free rollers,. and counterbalanced so that they can be very easily and rapidly raised.

The Richmond weir keeps up the Thames between Richmond and Teddington to half-tide level, in order to cover the banks in mid-channel, which used to be laid bare at low tide, owing to the considerable lowering of the low-water line of the river within the last sixty years, resulting from the removal of old London Bridge, old Westminster Bridge, and other obstructions, together with dredging in the river, facilitating the discharge of the ebb tide. Draw-doors, 12 feet high, close the three central openings, of 66 feet, of the arched double foot-bridge across the river, when the river falls below half-tide level; and being raised from the bridge as soon as the rising tide reaches that level, they leave the influx and efflux of the tide unimpeded up to Teddington as formerly above half-tide level. These draw-doors are counterbalanced, as at Ballinasloe, so as to be rapidly raised at the proper time; and they have to be raised and lowered every tide, except when a descending flood keeps up the river during low tide to the desired level. When raised, the doors are rotated into a horizontal position, so as to be placed out of sight between the footways, in order that they may not intercept or mar the view along the river.

Sluice-gates working on the same principle have been erected alongside the locks of the Manchester Ship-Canal, to regulate the discharge of the rivers Mersey and Irwell which join the canal, and also alongside the tidal locks at Eastham to regulate the level of the tidal reach of the canal. The largest work, however, of this kind on this canal is the draw-door weir or sluices erected, along the outer side of the canal, across the mouth of the river Weaver, with ten openings of 30 feet span, closed by counterbalanced draw-doors, or sluice-gates, capable of being raised 28 feet, which whilst maintaining the water-level in the canal, have to regulate the discharge of the Weaver so that its fresh water and tidal flow, into and out of the Mersey estuary, shall take place precisely as it did when the river flowed direct into the estuary.

This type of draw-door weir is specially well adapted for regulating the flow through it, and consequently the water-level above it, owing to the great facility with which the doors are raised and lowered, especially when counterbalanced; but as the discharge commences at the bottom of the weir, under a head of water, the apron of the weir requires protection against the powerful scour which takes place directly the door is lifted off the sill.

Segmental Gate Weir. A peculiar form of weir, erected in 1853, may be seen across a branch of the Seine near the centre of Paris. This weir of La Monnaie has four openings of $28\frac{2}{3}$ feet, separated by masonry piers, each closed by a wrought-iron cellular gate made in the form of a segment of a cylinder. The gate revolves on a horizontal axis, round pivots fixed in each pier, with which it is connected, at each side, by six radiating iron rods (Plate 3, Fig. 19). The gates descend into a curved depression in the apron when the weir is to be opened, and are counterbalanced by weights rising or falling in a hollow in the piers. They are by this means easily turned through a small arc, by chains worked from the piers, closing or opening the weir according as they are raised or lowered. This form of weir was adopted with the object of its being out of sight when the weir is open, and in order to dispense with any intermediate frames which would be liable to arrest floating rubbish, giving an unsightly appearance in the centre of the city. Each gate weighs 7½ tons; and the cost of each opening was £750. This weir, which retains the water up to about 8 feet above the apron, has worked satisfactorily, and has not at all impeded the passage of ice and floating substances.

Similar gates were erected about the same period for closing the weir erected across the Rosetta branch of the Nile for the purposes of irrigation [1]. This weir has sixty-one openings, each 16½ feet wide; and the gates, which, unlike the Paris weir, were raised out of the water for opening the weir, were intended to retain the water up to a height of 19 feet above the apron. The foundations, however, were subject to settlement in the alluvial ground; the water escaped through gratings under the gates, placed there in order to keep the gates clear of silt; and the water could never be retained to the desired height. Since 1884 the foundations of this weir have been repaired, and the gratings have been closed; and for the future this weir, and the Damietta weir, are to be controlled by iron draw-doors sliding against free rollers.

Concluding Remarks. The discharge of most navigable rivers generally suffices, even in the dry season, for supplying water for locking, so that the chief objects in designing river locks consist in adapting their sizes to the vessels using the navigation, and in expediting the passage through the locks, by providing intermediate gates or smaller locks for the small vessels, and by constructing ample sluice-ways fitted with revolving, or cylindrical sluice-gates, so that the lock may be

---

[1] 'Egyptian Irrigation.' W. Willcocks, p. 152, plate 20.

rapidly filled or emptied. The sills also of the locks in an important, or developing navigation, should be placed at a lower level than the existing available depth of the river, in order that any subsequent deepening of the channel may be effected without entailing the rebuilding of the locks.

Fixed weirs possess the advantages of requiring no looking after, and of rarely needing repairs ; but they more or less block up the natural channel of the river, and they only maintain the river to a certain level above, without at all regulating the discharge at any higher water-level. They should always be placed in a wide portion of a river, so as to ensure the greatest available capacity of discharge over their crest; but they are only properly suitable for torrential rivers flowing between steep banks. Even when forming irrigation dams across rivers which rise to a considerable height in the rainy season, and in their natural condition inundate large tracts of country, solid weirs are advantageously supplemented by sluice-ways which can be opened, in order that accumulations of deposit above the weir may be carried down.

Draw-door weirs form important adjuncts to fixed weirs, enabling the discharge to be regulated, and facilitating the passage of flood waters. They are, indeed, more complicated structures than fixed weirs, and necessitate superintendence for opening and closing them. No canalized river, however, subject to inundations, can be considered in a satisfactory state which is devoid of appliances for regulating its water-level, and of provision for the rapid discharge of floods at the weirs. Draw-door weirs afford a method of accomplishing these objects, but they generally only supplement fixed weirs, their sills are commonly raised above the bed of the river, and their doors are not lifted high enough to enable vessels to pass through the weir in flood-time. Accordingly, draw-door weirs, whilst furnishing a very valuable means of regulating a river, and capable of being placed under perfect control by the adoption of free rollers for reducing friction, generally extend across a portion only of the channel, and are not suited for navigable passes. Moreover, the frequent piers necessary for small draw-doors obstruct the channel; and the piers and overhead bridges required for large draw-doors involve a considerable expenditure. Consequently, draw-door weirs, though constituting a great advance on the rough simple system of fixed weirs, do not afford a perfect system of weirs for rivers in which it is important to leave the channel as open as possible in flood-time, and in which the navigation has to pass through the weir when the lock is submerged, as for instance on the Seine, the Main, the Meuse, and other large rivers.

CHAPTER VI.

MOVABLE WEIRS.

Definition; Classification. *Frame Weirs:*—Needle Weir, description, method of working, instances, improvements, advantages, limit of height; Sliding Panels, object, advantages, instances; Hinged Curtain, description and method of working, compared with panels; Suspended Frames, object, description of Poses weir, advantages, cost compared with other forms. *Shutter Weirs:*—Bear Trap, description, method of working; Thénard's Shutters, description, working, modifications adopted in India; Hydraulic Shutters, description and cost; Chanoine's Shutters, modifications and working, introduction of foot-bridge, cost, adoption in America, system of double grooves, advantages; Tilting Shutter, description. *Drum Weirs:* —Description, at Joinville, on the Main, at Charlottenburg, merits. General remarks, summary of respective types, and comparison of systems.

A MOVABLE weir is a barrier placed across a river for keeping up the water above it to any desired level, and capable of being so lowered on to the bed of the river, or removed, as to present no obstruction to the waterway in flood-time.

The primitive stanches used for flashing on rivers, as described in Chapter III, furnished the first rough types of movable weirs ; and the needle weir, which is a common form of movable weir at the present time, was first introduced on the river Yonne for facilitating navigation by means of flashes. The addition of locks at the weirs, for the purposes of a continuous navigation, obviated the necessity of a frequent opening of the weirs, or stanches, for the passage of vessels. Nevertheless, movable weirs, besides serving for regulating the discharge, are very valuable for enabling floods to pass down in an unrestricted channel, and are in many cases indispensable for the passage of vessels through them when the locks are submerged, and when the flood has not risen so high as to put a stop to navigation. A movable weir is generally divided by piers into two or more passes, one or more deep passes near the lock serving for the passage of vessels when the river is high, and one or more shallow passes in a shallow part of the river, or raised for economy and facility of working above the river-bed, for regulating the discharge and affording an outlet for small floods.

A variety of forms of movable weirs have been erected or experimented upon; and considerable modifications and improvements have been introduced from time to time; but movable weirs may be classified under three distinct types, namely, (1) Frame weirs closed by needles, panels, or a rolling-up curtain; (2) Shutter weirs; and (3) Drum weirs. All these types, with some of their varieties, are illustrated in section on Plate 4, drawn to the same scale for the purpose of comparison, the first type being represented by Figs. 7, 8, 9, and 11, the second type by Figs. 10, 12, 13, 14, and 15, and the third type by Fig. 16.

1. FRAME WEIRS.

Frame weirs consist of a series of iron frames placed in a row across a river, carrying a foot-bridge on the top, and hinged to the apron at the bottom, or suspended from an overhead bridge, against which the actual wooden barrier of needles, panels, or curtains rests. The needles, or square spars, can be removed successively by hand from the foot-bridge, the panels can be raised, or the curtains rolled up; and then the frames can be gradually lowered on to the bed of the river, or raised, in the most recent examples, some height out of the river, so as to leave the channel entirely open during floods for the discharge of the water and the passage of vessels.

Needle Weir. This form of frame weir derives its name (*barrage à aiguilles*) from the long square spars, rounded off at the top to form a handle, which, standing almost vertically in the water, side by side in a long row across the river, close the weir (Plate 4, Fig. 7). A row of light, wrought-iron frames, placed end on to the current at regular intervals apart, are connected and fastened together, when standing upright, by movable, horizontal, iron bars, and can be lowered flat down across the stream, being hinged to the apron of the weir. Each frame forms a trapezium braced horizontally and diagonally; the frame and braces being formed of bar iron, T irons, channel irons, or angle-irons. The lowest horizontal piece of the frame has iron pins at its extremities, which turn in cast-iron sockets fixed to the apron, and form the hinges of the frame. The horizontal pieces at the top of the frames support a foot-bridge of planks, or iron plates, on which the weir-keeper stands for removing or replacing the needles. The needles, when in place, rest against a sill on the apron at the bottom, and against the front horizontal bar connecting the frames at the top. The needles are almost always made square in section, as this form has proved more convenient in practice than oblong and hexagonal sections. The frames are generally placed at intervals of only between 3 and 4 feet, so that they may not be too heavy to be raised or lowered by the weir-keeper.

To open the weir, the needles are first raised one by one and removed, the movable bars connecting the frames are unfastened and withdrawn, and the frames are then gently lowered one after the other by the help of chains, the connecting bars being withdrawn and the foot-bridge taken up length by length as the frames are lowered, commencing from one end of the weir.

The weir is reinstated by raising each frame successively by means of the chains attached to them, the bars are fastened on each pair of frames directly they are raised, the foot-bridge is then relaid, and lastly the needles are put in place one by one.

Needle weirs were originally exclusively adopted for the weirs of the Lower Seine ; and some of the largest examples of this form of weir still exist on that part of the river, though several of the weirs have been reconstructed, and their form modified, to raise the water-levels of the respective reaches. Needle weirs have also been put across the regulating passes of the Upper Seine weirs, where the water-level of the reaches above them was raised by the latest improvement works. Needle weirs have, moreover, been erected on the Meuse and several other rivers ; and this system has been used for the principal portion of the weirs recently constructed for the canalization of the river Main. The sizes of the frames and needles depend upon the height of the water-level in the upper reach above the sill of the weir, which must be nearly level with the bed of the river in navigable passes, and also upon the head of water which may have to be retained at the weir. At Martot, the last weir on the Lower Seine (Plate 4, Figs. 2 and 7), the frames are 8 feet long at the base and 11 feet high, and weigh 467 lbs. each ; and the needles are 13 feet long and 3 inches square, the head of water retained occasionally reaching nearly 10 feet. The frames on the Belgic Meuse are 8⅓ feet long at the base, and 13 feet high, and weigh 798 lbs. ; and the needles are 4 inches broad, and 4 inches thick at the bottom, increasing to 4¾ inches at the centre, and weigh 55 lbs., reaching about the limit which a man can handle, so that contrivances have been devised for facilitating their manipulation. Thus on the Meuse and the Main, the bars supporting the needles along the top of the frames are hinged at one extremity to a frame, and the other end is held in position by a vertical spindle on the adjacent frame, so formed that by giving it a quarter of a revolution the end of the bar is released, and turning on its hinged end sets free the series of needles it supported, which float down stream and are fished out by means of a long rope passing through iron eyes fastened to each needle [1]. In the needle weirs on the Upper Seine, each needle has an iron hook attached near its handle, which encircles the bar against which it rests, so that the needle can rotate on the bar [2]. The needle is put in place by pushing it up stream across the foot-bridge till the hook catches the bar ; the needle then projecting out from the bridge, descends by its own weight, and, being pushed well down, the stream carries it into position, its end being arrested by the projecting sill. The needle is released by lifting it at its hook by a lever or a crab, till its end rises clear of the sill, when the current makes it rotate on the bar till its end floats in the current pointing down stream. By thus raising the needles successively, the weir is opened to any desired extent ; and the needles can be unhooked and removed by aid of a boat on the up-stream side, when the current through the weir has slackened.

Needle weirs are simple, comparatively cheap, and easily worked and repaired. Moreover, by removing a few needles the discharge can be regulated to any extent ; or sometimes the simple expedient of pushing out some of the needles at the top away from the frames, and keeping them in that position with blocks of wood, affords a sufficient outlet for the discharge, without the trouble of removing and replacing any of the needles. There is always a certain amount of leakage beween the needles ; but it rarely exceeds the discharge of the river that must be allowed to pass, if the needles are properly placed and in good condition. Means, however, have been adopted for reducing undue leakage through these weirs in dry weather, such as closing the interstices with hay, putting a row of covering needles in front, or stretching a tarred canvas curtain along the upper face of the needles.

[1] ' Canal, River, and other Works in France, Belgium, and Germany.' L. F. Vernon-Harcourt. Minutes of Proceedings Institution C. E. 1889, vol. xcvi, p. 186.
[2] Annales des Ponts et Chaussées, 1881 (2), p. 220, and 1883 (1), p. 622.

The limit of height to which these weirs can be carried has probably been reached at the needle weirs on the Lower Seine. the Belgic Meuse, and the Main. As the needles rest merely against the sill and the top connecting bar of the frames, their section has to be considerably increased when the height of the frames is raised. An attempt was made to reduce the strain on the needles in high weirs by introducing an intermediate supporting bar ; but this arrangement, whilst considerably relieving the needles, prevented the adoption of contrivances for releasing them, and also materially increased the difficulty of raising them, owing to their being pressed by the stream against the central bar. The rotating bar adopted on the Meuse and the Main, only releases the needles, and affords no assistance in replacing them. The hook on the back of the needles used in France is of considerable assistance in replacing the needles, as soon as the hook is in position ; but though the system has been tried with heavy needles, the process of hooking on must always be somewhat difficult; and hitherto, in France, the size of the needles has been limited to 3 inches square, which is too slight for high weirs.

The increase in the navigable depth of the Lower Seine, from 6½ feet to 10½ feet, has necessitated the raising of several of the weirs, as well as the construction of additional weirs, retaining a greater head of water than the original needle weirs. As, however, the suitable limit of size for the needles was considered to have been already reached on the Lower Seine, it became necessary to devise some other form of barrier for closing the higher weirs, which could be more easily handled than needles.

Sliding Panels on Frame Weir As the movable frames of the needle weirs are only between 3 and 4 feet apart, the distance between the points of support of a transverse barrier resting against the frames is much less than the distance between the sill of the weir and the upper bar against which the needles rest, which, moreover, increases with the height of the weir. Accordingly, the thickness of a transverse barrier resting against two adjacent frames may be considerably less than the thickness of the needles ; and, moreover, its thickness may be reduced with the diminution in the water pressure from the bottom towards the surface. Consequently, by adopting a series of wooden panels sliding between, and resting against each pair of frames, for closing the space between them, instead of needles, the thickness of the barrier can be considerably reduced without diminishing its strength, and proportioned to the variation in pressure.

The weight of each panel is regulated by its height, which is made dependent on the limit of weight that can be easily lifted by a crab on the foot-bridge, winding up a chain hooked on to an iron loop attached to each panel. This system was experimented on by Mr. Boulé at Port-à-l'Anglais weir [1] on the Upper Seine in 1874, and introduced by him at Suresnes weir on the Lower Seine on its reconstruction in 1880-85 (Plate 4, Fig. 8) [2]. The navigable pass of Suresnes weir has frames about 20 feet high ; and the opening between each pair of frames is closed by four panels sliding in grooves on the up-stream faces of the frames, each panel having a height of 4¼ feet and a width of about 4 feet. The panels are pushed down into their places by a pole, and are raised by the crab travelling along the foot-bridge on the top of the frames, which lifts each panel in succession by means of a hook and chain. The number and length of the joints in the panel weir are much less than with needles, so that the leakage through the weir is considerably reduced ; and the foot-bridge can be raised high enough to be secure from submergence, without increasing the height of the panels. The discharge of the weir can be regulated by removing some of the upper panels ; and this system has the advantage,

---

[1] Annales des Ponts et Chaussées, 1876 (1), p. 320.

[2] 'Les Écluses de Suresnes et de Bougival,' A. Boulé. Mémoires de la Société des Ingénieurs Civils de Paris, 1884 (2), p. 246.

over draw-doors and needles, of not producing a scour along
the apron on partially opening the weir, since the discharge
takes place over the top of the panels still remaining in place.
The system of superposed panels for closing a frame weir
is simple, economical, and easily worked; and the height
which such a weir could attain is only restricted by the limit
of weight imposed on the frames by the necessity of their
being able to be raised without difficulty from the bed of
the river, when the weir has to be reinstated on the approach
of the summer season.

Before this form of frame weir was employed at the reconstructed Suresnes weir, completed in 1885, it had been
adopted with success for the regulating portion of the Mulatière
weir across the Saône at its confluence with the Rhone below
Lyons: and it was also employed in 1876 for six weirs on the
river Moskowa between Moscow and Kolumna, but in this
case the height of the panels was made only 1 foot, so that
they could be easily raised by a boat-hook.

Hinged Curtain on Frame Weir. Another form of transverse barrier, supported by the movable frames, was designed
by Mr. Caméré for closing the openings between the frames
of Port Villez weir, the first of the new weirs erected on the
Lower Seine for increasing the navigable depth to 10½ feet,
and which was opened in 1880. It resembles in principle the
panel system, in being composed of transverse pieces of wood
resting on the frames, and proportioned in thickness to the
depth of water; but it forms a continuous curtain hanging
down the up-stream side of a pair of frames, consisting of a series
of strips of wood joined by hinges, which covers the opening
from top to bottom (Plate 4, Fig. 9). The curtain is rolled
up, from the bottom, by an endless chain turning round a crab
travelling along the foot-bridge on the top of the frames.
The rolling up or unrolling of the curtain can be stopped at
any point, so that the extent to which the weir is opened or
closed can be regulated with great nicety. As, however, this

form of weir is opened from the bottom, its apron is exposed
to a powerful scour when the raising of the curtain is commenced.   The weir at Port Villez, 700 feet long, is placed alongside the lock which is close to the left bank, the two navigable
passes placed nearest the lock having frames across them,
18 feet high and 3 feet 7¼ inches apart, each weighing 37 cwts.,
and capable of being raised or lowered in ten minutes[1]. Frames
of half the height are placed across the two shallow regulating
passes adjoining the right bank of the river ; and the foot-bridge
is only 1¾ feet above the upper water-level.   The curtains
which close the whole of this weir have worked satisfactorily,
and have proved watertight ; and similar curtains have been
adopted at the other more recently constructed weirs on the
Lower Seine (Plate 4, Fig. 11), with the exception of Suresnes
weir, where however, in one of the passes, panels and curtains
have been placed side by side to test their relative durability.

The hinged curtain possesses the advantages of being readily
raised or lowered to any extent, and of being capable of
exactly adjusting the discharge through the weir ; whereas
the panels are not so readily raised when the water is flowing
over them.   On the other hand, the panels are cheaper and
less complicated, and will probably prove more durable ; and
with them, the apron is not exposed to scour.

Suspended Frames with Curtain.   Though the height of
the barrier of a frame weir could be considerably extended by
the substitution of panels or rolling-up curtains for needles,
the size of the frames which can be conveniently raised and
lowered has been practically reached at Port Villez and
Suresnes weirs.   By suspending the frames from above, however, instead of hinging them on the apron at the bottom of
the river, it is possible to raise all the movable parts of the
weir out of water in flood-time, and to render the frame
weir applicable to greater heights (Plate 4, Fig. 11).   This

---

[1] Minutes of Proceedings Institution C. E., vol. lx, plate 4 ; and vol. lxxxiv,
plate 3, fig. 7.

principle was proposed many years ago by Mr. Tavernier for weirs on the Rhone, where ordinary frame weirs, with the frames lowered in flood-time on to the bed of the river, would be impracticable, owing to the large quantities of gravel and shingle carried down by the rapid current, which would bury the recumbent frames. The system which has been carried out at Poses and Port-Mort weirs on the Lower Seine[1], consists of a series of frames suspended vertically from a wide overhead foot-bridge, and hinged so that they are capable of being raised to a horizontal position under the bridge, sufficiently high at the navigable passes to afford the requisite headway for vessels underneath, so long as the river does not rise above the highest navigable level. The frames, when lowered, rest at the bottom against a sill; and the spaces between them are closed by hinged curtains, rolled up or let down by the weir-keeper from a small foot-bridge formed of brackets hinged to the back of each frame.

Poses weir is separated from the locks by an island, and is divided by masonry piers into seven passes (Plate 4, Fig. 4). The two passes adjoining the left bank are navigable passes, each 106½ feet wide, with their sills 16½ feet below the upper water-level; and they leave a headway of 17¾ feet above the highest navigable flood-level when the frames are raised, which is 1⅜ feet higher than the ordinary upper water-level, but 3⅛ feet below the maximum height reached by the great flood of March 1876 (Plate 4, Fig. 11). The five other passes are each 99 feet wide; and the girders across them, carrying the overhead foot-bridge, are placed at a lower level than across the navigable passes, as no headway is needed, so that the platform on the top of them is at the same level as the platform placed between the girders across the navigable passes, making the footway level throughout. Each frame consists of four long ribs braced together, across which

---

[1] 'The River Seine,' L. F. Vernon-Harcourt. Minutes of Proceedings Institution C. E., 1886, vol. lxxxiv, pp. 234 and 236; and plate 3, figs. 1, 3, and 4.

a curtain, 7½ feet wide, extends, an interval of 1½ inches being left between the curtains of adjoining frames for clearance. The rolled-up curtain rests on the frame when the frame is raised, which is effected by means of a winch and chain from the foot-bridge; and the curtain is only removed for repairs. The rolling-up or unrolling of a curtain takes fifteen minutes; and a frame is raised in twenty minutes, or lowered in ten minutes; so that a pass can be fully opened in five hours.

This system of frame weir is very readily opened or closed, and secures the frames from the injuries to which they are exposed when laid lapping over one another on the river-bed, and left under water during the winter, and also from the strains to which they are subjected in the process of lowering them into the river, or raising them from the river-bed. Moreover, the suspended frames have only to be made strong enough to bear the strain of being raised, and the water pressure against the curtain when the weir is closed. On the other hand, the system involves the costly addition of a wide foot-bridge spanning the openings, and also long frames and high piers at the navigable passes on account of the headway required. Nevertheless, the cost of Poses weir has been stated by Mr. Caméré, the engineer who designed it, to have been £151 5s. per lineal foot; whereas the cost of Port Villez weir amounted to £163 7s. per lineal foot [1], though more than half the cost in both these weirs was due to the very deep foundations necessitated to prevent filtration under the weirs through the alluvial subsoil. Probably, however, the cost of Port Villez weir was considerably increased by experimental systems tried there, for on comparing the section of Poses weir with the sections of Suresnes and Port Villez weirs, drawn to the same scale, (Plate 4, Figs. 8, 9, and 11), it is impossible to avoid the conclusion that, under

[1] At Suresnes, the weir across the neighbouring pass, with frames 19¾ feet high, cost £148 per lineal foot; and the regulating weir, with frames 13¾ feet high, cost £93 per lineal foot.

similar conditions, for heights not exceeding those of these weirs, the cost of suspended frames, with their wide overhead foot-bridge, must be more costly than simple frames. The Martot needle weir (Plate 4, Fig. 7) cost only £49 per lineal foot, though it is one of the largest examples of ordinary frame weirs. The chief merits of the suspended frame weir are ease and security in working, exemption from injurious influences, facility of maintenance, and its applicability to greater heights than those reached at Suresnes and Port Villez.

2. SHUTTER WEIRS.

Shutter weirs are composed of a series of hinged panels, gates, or shutters, which, when raised, form a barrier across the river; and the weir is opened by making the shutters revolve on a horizontal axis, and fall flat on the bed of the river. Various methods have been devised for raising and supporting the shutters, which are generally made of wood and occasionally of iron; and in some types the shutter revolves on an axis at the bottom (Plate 4, Figs. 12, 13, and 15), though in the most common form of shutter weir, the axis is nearly central (Plate 4, Fig. 10).

Bear-Trap Weir. A shutter weir, called a bear trap, was erected, in 1818, on the Lehigh river in Pennsylvania. It was formed by two wooden gates revolving on horizontal axes on the apron, the up-stream gate pointing down stream, and the down-stream gate pointing up stream. When raised, the up-stream gate was supported on the top edge of the down-stream gate. This weir is not now in existence; but the type has been reproduced, with improvements, in France, at the weir of La Neuville-au-Pont on the river Marne (Plate 4, Fig. 15). The pass in which the gates are placed, is 29½ feet wide. Culverts have been constructed in the side walls, with outlets into the upper and lower pools; and a side passage from each of the culverts opens into the space under the gates. For closing the weir, a small shutter, revolving on

a horizontal axis on the apron above the upper gate, is
raised, rising with the stream. This shutter stops the flow
of water through the weir, and removes the pressure from
the recumbent gates. Then by opening the upper sluice-
gates, the lower ones being closed, the water flows from
the upper pool into the space under the gates and raises
them. The weir is opened in three minutes, by releasing the
fastenings of the gates, closing the upper sluice-gates, and
opening the lower ones, so that the water under the gates
flows away into the lower pool, and the gates fall down on
losing their support. At this weir, the water in the lower
pool is liable to fall too low for navigation before the requisite
head of two feet is obtained in the upper pool for closing the
weir, and consequently the gates are raised by chains.

This type of weir requires a greater pressure of water for
working it than is available at La Neuville; and it is costly
with its large gates, long side walls, and culverts, and is not
suitable for wide passes, so that other forms of shutter weirs
have been since preferred. A drift pass, however, has been
recently constructed on this principle in the Davis Island
weir on the Ohio, 52 feet in width, to afford a convenient
outlet for the large quantities of drift brought down by the
river, as the bear trap is more readily raised again after
the discharge of the drift than one of the shutters of the weir
during the low stage of the river.

Thénard's Shutter Weir. The earliest type of shutter weir
was erected across a shallow pass on the river Orb in the
eighteenth century. It consisted of a shutter, turning on
a horizontal axis at the bottom, supported by a prop when
raised against the stream, and falling flat on the apron
when the prop was drawn aside; but it was difficult to raise
the shutter against a head of water. Accordingly, when
Mr. Thénard erected similar shutters across three shallow
passes on the river Isle, between 1832 and 1837, he placed
a second set of shutters on the up-stream side of the weir,

which, rising with the stream, were retained in an upright position by chains, and stopping the flow of water through the weir, enabled the down-stream shutters to be easily raised and propped up[1]. Valves were then opened in the up-stream shutters ; and the level of the water being equalized on both sides of them, they could be lowered and secured to the apron, the other shutters remaining upright and forming the actual weir. The weir was readily opened by drawing aside the prop of the down-stream shutter. The St. Antoine weir, erected in 1843, having seven shutters 5½ feet high and 4 feet wide, was an extension of the same system. A narrow footway was constructed along the top of the up-stream shutters, from which the weir-keeper could lift the others.

This form of shutter weir has not been further employed in France, for it was found that the river had to drop to a lower level than suitable for navigation before the weir could be raised. Moreover, the up-stream shutters, in rising with the stream, were liable to come up too rapidly and tear away the fastenings of their chains.

Indian Shutter Weir. A similar type of shutter weir has been employed on a large scale for closing the openings in some Indian irrigation dams. In the Mahanuddee dam, there are ten openings, 50 feet in width, each closed by seven pairs of shutters, the down-stream shutters forming the actual weir being 9 feet high[2]. The sixty-six openings in the Sone dam, 20½ feet wide, are each closed by a single pair of shutters, the down-stream shutter being 9½ feet high (Plate 4, Fig. 13). As the up-stream shutters of the French weirs were liable to injury, though only raised when the discharge of the river was becoming small, the larger shutters of the Indian weirs, at the head of irrigation canals, are naturally exposed to considerably greater strains, as they have to be

---

[1] Annales des Ponts et Chaussées, 1841 (2), p. 45.

[2] 'Movable Dams in Indian Weirs,' R. B. Buckley. Minutes of Proceedings Institution C. E., 1880, vol. lx, p. 44.

raised when a sudden fresh in the river threatens to discharge too rapid a current into the canal, leading to a frequent breakage of the chains or their fastenings. To regulate the rising of the large shutters in the Sone weir, Mr. Fouracres fitted a hydraulic brake to their up-stream side, which consists of a piston drawn along a cylinder full of water, with small orifices in the side. As the shutter rises with the current, the motion of the piston is checked by the cushion of water in the cylinder, resulting from the inability of the water to escape rapidly through the small orifices. Moreover, as the piston travels along the cylinder, the number of orifices available for the escape of the water is reduced; so that whilst the tension on the piston rod is increased as the shutter rises, the checking force is also augmented.

Hydraulic Shutter Weir. In a shutter weir erected across a shallow pass at Brûlée Island, on the Yonne, in 1870–72, Mr. Girard dispensed with protecting shutters, and raised the shutters against the stream by hydraulic power exerted by a press under each shutter (Plate 4, Fig. 12)[1]. A piston, working in the cylinder of the press, is fastened at its upper end to a cross-head carrying three connecting rods by which it transmits its motion to the shutter, the rods being fastened to an axis placed along the central line of the underside of the shutter. The water pressure is supplied to the presses from an accumulator on the bank, the water being pumped into the accumulator by a turbine which is turned by the fall of water at the weir. The seven shutters of this weir, each 11½ feet long and 6 feet high, can be raised in five minutes. The working of this weir has proved quite satisfactory; and the large cost of this type of weir, amounting to £36 6s. per lineal foot, about three times the cost of a needle weir, or ordinary shutter weir of similar moderate height, is the only reason which has prevented the extension of the system.

Chanoine Shutter Weir. At a weir erected at Conflans on

---

[1] Annales des Ponts et Chaussées 1873 (2), p. 360.

the Upper Seine, in 1858, Mr. Chanoine completely transformed the arrangement of the shutters, so as to render the system suitable for navigable passes as well as shallow passes [1]. The axis on which the shutter turns is placed horizontally, a little above the centre of pressure, on the down-stream side of the shutter, and is supported on a wrought-iron tressel hinged to the apron of the weir (Plate 4, Fig. 10). When the weir is closed, the tressel stands upright; and the shutter, having a slight inclination down-stream at the top, abuts at the bottom against a projecting sill on the apron, and is supported in that position by a wrought-iron prop which is connected with the shutter and tressel, and turns, together with the tressel, on the same axis as the shutter. The prop, when supporting the shutter, is inclined at an angle of about 45°; and its lower extremity rests in a cast-iron shoe, from which it can be released by a sideways pull by a bar with projecting teeth, called the tripping bar, which is laid on the apron and worked by a man from the bank. When the prop has been drawn clear of the shoe, it slides down an iron groove on the apron; and the tressel and shutter, being unsupported, are pushed over by the stream, and fall, over the prop, flat down on a recessed part of the apron.

The shutter is raised, in a horizontal position, by a long hook from a boat, or from a foot-bridge above the weir, by a chain attached to its lower extremity. As soon as the tressel has been raised, and the prop is in the shoe, the lower extremity of the shutter is released, and it closes, either under the pressure of the water alone, or, if necessary, by the aid of a push at the bottom, or by a pull from a chain attached to the top.

The original intention was that the small shutters across the shallow passes should automatically regulate the water-level of the river, their axis of rotation being so adjusted, near a third of the height up, that a rise of the river above the

[1] Annales des Ponts et Chaussées, 1859 (2), p. 197.

desired level would make the shutters rotate and open the weir, till the drop in the water-level produced by the increased discharge reversed the balance of pressure on the upper and lower portions of the shutters, and closed the weir. All the weirs on the Upper Seine were constructed on this principle; but it was found that, though the regulating pass of the weir opened when the river attained a certain level above it, the increase in the discharge was too suddenly produced; and, moreover, the river fell too low before the shutters righted themselves and closed the weir[1]. Accordingly when most of the Upper Seine weirs were raised for increasing the navigable depth, needle weirs were substituted for shutter weirs across the shallow regulating passes, it having been found previously necessary to add a foot-bridge on movable frames to the regulating shutters, in order to control their movements by means of chains[2].

The axis of the large shutters across the navigable passes is placed sufficiently above the centre of pressure to prevent their opening with a rise of the river, even when the water rises above its ordinary level on the down-stream side; but provision is generally made for increasing the discharge along these passes when the river rises, by inserting a butterfly valve, resembling a diminutive shutter, in the upper panel of the shutter, turning on a horizontal axis near the centre of pressure so as to open when the water-level attains a few inches above the top of the shutter. These valves also, which in the Upper Seine weirs are $3\frac{1}{8}$ feet high by $2\frac{1}{8}$ feet wide, facilitate the raising of the weir. The shutters across the navigable passes on the Upper Seine, $11\frac{1}{2}$ feet high and 4 feet 1 inch wide, are raised from a special boat, according to the original design; but a foot-bridge on movable frames has in many cases been placed alongside the shutters across

---

[1] Annales des Ponts et Chaussées, 1861 (2), p. 209; 1868 (1), p. 282, and (2), pp. 50, and 366; and 1873 (2), p. 159.

[2] Ibid., 1883 (1), p. 622.

navigable passes to facilitate their working, as for instance on the Marne, the Saône, and the Ohio. Some large shutters were placed across a deep pass of Port-à-l'Anglais weir on the Upper Seine, in 1870, which had been formed through a portion of the shallow pass in order to supply the place of the lock during repairs[1]; these shutters can retain a height of 12 feet of water above the sill of the weir, they are 13 feet high and 3¼ feet wide, and they are raised or lowered from a foot-bridge on movable frames 15¾ feet high (Plate 4, Fig. 10). The cost of this high weir, with its foot-bridge, was £67 12s. per lineal foot; whereas the average cost of the weirs, as originally constructed across the navigable passes on the Upper Seine, 10 feet high, was £37 3s. per lineal foot, and across the shallow regulating passes 6½ feet high, together with a foot-bridge, was £21 16s. per lineal foot. The shutter weirs erected across some navigable passes on the Marne, 9½ feet high, together with the foot-bridge on movable frames, cost £51 4s. per lineal foot.

The Chanoine system of shutter has been adopted for weirs for improving the navigation on the Great Kanawha and Ohio rivers in the United States, three movable weirs and two fixed weirs having been erected in 1880–87 on the Great Kanawha river above and below Charleston, for securing a navigable depth of not less than 6 feet throughout the year, and one movable weir having been constructed at Davis Island on the Ohio, five miles below Pittsburgh[2]. The weirs on the Great Kanawha river differ from the system described above, in substituting iron rectangular frames hinged to the apron for the tressels, in which the shutters revolve; but the weir on the Ohio was made on the Chanoine model, with the addition of a foot-bridge on movable frames across the navigable pass, 559 feet wide, to facilitate its working, as well as along the three shallow passes having a total width of 664 feet.

[1] Annales des Ponts et Chaussées, 1873 (2), p. 98.
[2] 'Reports of the Chief of Engineers, U.S.A.' 1881, 1889, and 1892.

The shutters across the navigable pass at Brownstown on the Great Kanawha river, nine miles above Charleston, are 13½ feet high and 3⅔ feet wide, slightly larger therefore than the large shutters at Port-à-l'Anglais weir. Two more weirs are in course of construction below Charleston, and several others are in contemplation lower down. It has been found necessary to dispense with the foot-bridge at the weir on the Ohio, as the frames were liable to be fractured by accumulations of drift and floating ice before they could be lowered; and the weir is now raised and lowered from a boat. The cost of this weir, together with a lock, was about £191,000. Though this weir is very much hampered by the large quantities of drift, and the frequent inability to lower it before a flood has actually arrived, for fear of stranding fleets of coal vessels which congregate above it, its utility is so great for the navigation that the construction of four more similar weirs has been recommended, to extend the slack-water navigation 25 miles below Pittsburgh [1]. The weirs on these rivers have to be lowered and raised several times in the year, to provide for the passing down of sudden floods, and also in order to afford a passage for the vessels, too numerous to be accommodated by the lock, as soon as the river below is sufficiently high.

The tripping bar enables the shutters to be lowered quickly from the bank or from a pier; and in order to prevent too great a strain being thrown upon the bar, its projections for drawing aside the prop are so spaced that on first moving the bar, the shutters are tripped singly, on account of the pressure of water against them; and then as the pressure becomes less by the opening of the weir, the shutters are tripped in pairs, and lastly three, or even four together. In spite, however, of various precautions, stones are liable to impede the action of the tripping bar, and some teeth are occasionally broken off, delaying the lowering of the shutters. To obviate this difficulty,

[1] Report of the Chief of Engineers, U. S. A., for 1889,' part 3, p. 1869.

Mr. Pasqueau devised a special form of double groove for the foot of the prop to travel in, which was adopted at the Mulatière weir at the confluence of the Saône and the Rhone just below Lyons[1]. When the shutter is raised, the prop is drawn along the groove into its shoe in the ordinary way; but for lowering the shutter, the prop is drawn a little further up by a pull on the shutter, till it drops into a second groove, down which it slides directly the shutter is released, causing the shutter to descend on to the apron. Thus by standing on the foot-bridge and pulling upon a chain attached to the lower end of the shutter, the shutter is raised, and the prop brought into position; and it is lowered again by a little further pull up-stream, without the aid of a tripping bar. This system was followed at Davis Island weir, where the drift and gravel brought down by the river seemed to preclude the use of a tripping bar, which, moreover, could not have been applied to a pass of the width of 559 feet without intermediate piers; and it is also being tried at the weirs on the Great Kanawha river. The partial destruction of the foot-bridge at Davis Island weir has led to the adoption of a pusher, worked from a boat below the weir, to give the necessary movement of the prop up-stream for effecting the lowering of the weir. At the Mulatière weir, which was first worked in 1882, the size of the iron shutters, 14⅓ feet high and 4½ feet wide, together with the great width of 340 feet of the navigable pass, and its peculiar position at the junction of two rivers differing greatly in the periods of their floods, rendered it expedient to dispense with a tripping bar. The system of two grooves has worked there with perfect success, enabling the sixty-nine shutters to be easily raised and gently lowered from the foot-bridge in any order; and when any débris accumulates under a shutter, it is readily scoured away by lowering and raising the shutter, an operation which would be impracticable with a tripping bar.

[1] 'Barrage de la Mulatière, Notice sur le nouveau système de Hausses.' A. Pasqueau.

The shutter weir possesses the advantage over the frame weir of being capable of more rapid lowering, especially by the use of a tripping bar, which has rendered it peculiarly suitable for the rapid and frequent floods of the Great Kanawha and Ohio rivers. It is also superior to the needle weir in allowing a moderate increase in the discharge to pass over its crest, and also in providing for the passage of small floods by the ready opening of the butterfly valves, which in the Mulatière weir, attain the dimensions of 5 feet high by 3 feet wide. The irregularity, however, in the automatic closing of the regulating shutters across the shallow passes, leading to the addition of a foot-bridge, and the adoption of a foot-bridge in many cases for working the shutters of the navigable passes, render the system more complicated and costly than the frame weir.

Tilting Shutter Weir. Another form of shutter weir was erected, in 1882, across the River Irwell at Thostlenest, close to Manchester, by Mr. Wiswall, consisting of fourteen wooden shutters, 12 feet long and from 9 to 10 feet wide, turning on a nearly central axis supported on a fixed open framework extending across the weir [1] (Plate 4, Fig. 14). When the weir is closed, the shutters are inclined down-stream at an angle of 35° to the vertical, abutting at the base against the flat apron of the weir; and they were designed to tip over and assume a horizontal position when the river rises 2¾ feet above their crest, thus automatically opening the weir. The weir is, however, usually opened by chains fastened to the centre of the bottom of each shutter, passing round guiding pulleys and worked by two crabs, one on each side of the river, each controlling the seven chains of half the shutters, as the automatic opening is too sudden in its operation, and is liable to cause undue erosion of the river banks. The shutters are retained in their horizontal position by ratchets; and when released on the lowering of the water-level, they are closed by the

[1] 'The Engineer.' Sept. 1, 1882.

stream. The shutters themselves when raised offer hardly any impediment to the discharge, as they are protected by the upper transverse beam of the framework ; but the permanent framework reduces the sectional area of the waterway through the weir by rather more than one fifth. This form of shutter weir, accordingly, though regulating the discharge with precision, offers some impediment to the flow when open ; and, moreover, it would not be suitable for very wide openings on account of the multiplication of the controlling chains, and it is not applicable to navigable passes.

3. DRUM WEIR.

The drum weir was designed by Mr. Desfontaines for the shallow regulating passes of twelve weirs on the River Marne, erected between 1857 and 1867, to which river this type of weir was for many years restricted[1]. It consists of an upper, and an under iron paddle, capable of making a quarter of a revolution round a central horizontal axis. The upper paddle forms the weir, which it closes when upright ; and the under one, slightly larger than the other, revolves in a closed recess, below the sill of the weir, having the form of a quarter of a cylinder. which has led to the name drum being given to this type of weir (Plate 4, Fig. 16). The upper paddle is straight ; but the under paddle is curved in such a manner, near its junction with the axis, that a space is left between its upper face and the top of the drum, when both paddles are horizontal. through which water can be admitted so as to press upon its upper face. A similar space is left, by a prolongation of the drum, between the down-stream face of the under paddle and the vertical wall of the drum when both paddles are vertical.

Culverts are provided in the abutment of the weir, com-

[1] Annales des Ponts et Chaussées, 1868 (2), p. 482.

municating with the upper and lower reaches of the river
and the drum ; and the flow through these culverts is controlled by a see-saw arrangement of sluice-gates, whereby
communication can be opened and closed between the upper
or lower part of the drum and the upper or lower reach,
thus producing or removing a pressure of water on the upper or
lower face of the under paddle. and thereby closing or opening
the weir to any extent with great facility and precision.

The largest weir of this type in France was erected across
the regulating pass of the last weir on the Marne at Joinville,
in 1867, which consists of forty-two upper paddles forming
the actual weir, 3½ feet high and 4⅔ feet wide ; and this weir
can be completely opened or closed in three or four minutes
by one man.

Bigger drum weirs, precisely similar in type, have been
placed across the timber passes of the four weirs constructed
for canalizing the Main in 1883–86 (Plate 4, Fig. 5). The
pass at each weir, 39½ feet wide, is closed by a single paddle
retaining a head of 5 feet 7 inches above the sill of the weir.
In this case, the upper paddle constituting the weir is slightly
inclined down-stream when closed, so that when it is lowered
into a recess in the apron in opening the weir, the straight
under paddle, which is vertical when the weir is closed, may
leave a sufficient interval between its upper face and the top
of the drum to admit the water from the upper reach when the
weir is to be closed [1]. These weirs serve to regulate the water-
level in the reach above them, as well as for the passage of
timber ; and they are readily closed against the full discharge
of the river rushing through the pass, when the rest of the
weir is closed.

Another still bigger drum weir has been placed across the
navigable pass of the Charlottenburg weir on the River Spree
(Plate 4, Fig. 16), retaining a depth of 9⅛ feet above the sill

[1] 'Some Canal, River, and other Works in France, Belgium, and Germany,' L. F.
Vernon-Harcourt. Minutes of Proceedings Institution C.E., 1889, vol. xcvi, p. 191.

of the weir, which is 2¼ feet above the bed of the river, the normal difference in the water-level of the upper and lower reaches being 4 feet[1]. The width of the single upper paddle closing this weir is 32¾ feet; whilst the arrangement of the upper and under paddles and the drum resemble very closely the drum weirs on the Marne on an enlarged scale.

The great merit of the drum weir is the perfect control maintained over its movements, and the facility with which it is raised against a rapid current. As a hydraulic contrivance it is, indeed, superior to any other form of movable weir; and its sole drawback is that the drum containing the lower paddle, extending to a greater depth below the sill than the actual height of the weir closed by the upper paddle, involves costly foundations, rendering this type of weir expensive and preventing its ready adoption for deep passes. In fact, except at Charlottenburg, it has only been applied to shallow passes; and even at Charlottenburg the pass is moderate in depth. Thus the cost of the drum weir at Joinville was £27 12s. per lineal foot for an available height of weir of only 3½ feet; and it reached £144 16s. per lineal foot at the Charlottenburg navigable pass. Nevertheless, the great facility and rapidity of working, and the very small cost of maintenance, compensate to some extent for the large first cost; whilst the preciseness with which the drum weir can regulate the discharge, and its capability of being raised against a rapid current, render it a most important adjunct to a movable weir under certain conditions.

General Remarks on Movable Weirs. Though various kinds of movable weirs have been experimented on, and different forms of shutter weirs have been occasionally constructed, there are only three types that can be said to have entered into ordinary use; namely, the Frame Weir with modifications merely of the barrier adopted, the Chanoine

[1] Minutes of Proceedings Institution C.E., vol. xcvi, p. 192; and Zeitschrift für Bauwesen, 1886, p. 338.

Shutter Weir, and the Drum Weir; and of these, the drum
weir has been somewhat limited in its application.

The frame weir with needles, the simplest form of movable
weir, was the first type at all generally employed; and though
it was introduced more than half a century ago, it is still in
very common use in France and in Belgium, and was selected
in recent years for the weirs on the Main. The only objections
that can be urged against it are, its incapacity for regulating
sudden rises of a river, which may submerge the foot-bridge
before an adequate waterway can be provided by removing
a sufficient number of needles; and the limit to its height
imposed by the increasing weight of the needles, and of the
frames, necessitated by augmentations in height. The first
difficulty was provided against, at the needle weirs on the
Meuse below Namur, by the addition of a long solid weir
parallel to the stream, over which any increased discharge
could flow without unduly raising the water-level till the
weir-keeper could release some needles; and it has been
more effectually obviated by the drum weirs on the Main.
The second objection has been met by introducing panels or
curtains in the place of needles, and lastly by suspending the
frames from an overhead bridge, which renders the frame
weir available for any height of weir likely to be required in
a river.

The Chanoine shutter weir, though introduced sixteen years
after the Poirée needle weir, was fortunate in coming into
use at a period when navigation by aid of movable weirs was
being rapidly extended in France; and it was largely employed on the rivers of the Seine basin above Paris. In its
primitive form, by dispensing with a foot-bridge and providing an automatic regulating pass, it naturally became a
powerful rival of the needle weir, especially as by means of a
tripping bar its opening could be rapidly accomplished. This
system was specially well adapted for navigation by flashes,
which was still to some extent in use on its introduction.

When, however, continuous navigation by aid of locks became general, the delay in the automatic closing of the regulating shutters proved inconvenient; and the addition of a foot-bridge to remedy this defect rendered the system more complicated and costly than the needle weir, which has, accordingly, in some cases replaced the shutter weir when improvements for navigation necessitated the reconstruction of the weir. The insertion of a butterfly valve in the upper panel of the shutters has materially facilitated the regulation of the discharge, and the raising of the shutters; whilst the provision of a double groove for the prop to move along has enabled the tripping bar, with its risks of breakages, to be dispensed with, and the shutters to be lowered without any shock, though involving generally the addition of a foot-bridge. The Chanoine shutter weir, like the needle and drum weirs, was first employed in the Seine basin; but it has also been adopted on the Loire, the Saône, the Meuse above Namur, and in the United States. After due consideration of the different types of movable weirs in use, the Chanoine shutter weir was selected for the Great Kanawha and Ohio rivers, on account of the rapidity with which it could be lowered, an operation which is necessary on those rivers several times in the year, the tripping bar arrangement being used in the one case, and the double groove in the other. The partial destruction, indeed, of the foot-bridge of the Davis Island weir, by the accumulated drift coming down the river, shows that the frame weir would not be suitable under such conditions. The shutter weir, accordingly, in its primitive form without a foot-bridge, possesses distinct advantages over the frame weir in certain circumstances, and even with a foot-bridge in special cases such as at the Mulatière navigable pass, owing to the facility with which it can be lowered; but under ordinary conditions, the frame weir with rolling-up curtains or sliding panels appears preferable to the shutter weir.

The drum weir is unquestionably the most perfect type of
movable weir, owing to the perfect ease, regularity, and
rapidity with which it can be raised or lowered, giving the
weir-keeper absolute control of the discharge by merely
moving the handle regulating the sluice-gates, without the
need of a foot-bridge or other accessories. Its sole defect is
its cost, resulting from the deep foundations required by the
drum below the sill of the weir, which for some years limited
its application to the shallow regulating passes of the Marne
weirs. The extension, however, of the system to the timber
passes on the Main, where it performs the double office of regulating the discharge and providing at any time a passage for
rafts of timber, and still more the drum weir at Charlottenburg
across the navigable pass, show that the system can be
applied to all the purposes of a movable weir. Moreover,
the success which has attended the working of these enlarged
drum weirs, suggests the idea that this system might have
been as applicable to the improvement of the Great Kanawha
and Ohio rivers as the shutter weir, considering the moderate
depth of water to be retained and the obstacles to the
maintenance of a foot-bridge, and that the larger cost
might have been compensated for by the safety and rapidity
of working, the freedom from accident, the exemption from
heavy costs for maintenance, and the facility with which the
floating drift could be passed down.

Each of the three systems has its special advantages ; but
it appears probable that, in Europe, the frame weir, and
perhaps the drum weir, especially for regulating weirs, will be
more extensively used in the future than the shutter weir.

CHAPTER VII.

PREDICTION OF FLOODS; AND PROTECTION FROM INUNDATIONS.

Causes of Floods. *Prediction of Floods:*—by Rainfall Returns; various circumstances producing Flood Rise; Observations of Rise of Tributaries; accuracy attained. *Protection from Inundations:*—necessity for, in certain cases; Methods adopted; Extension of Vegetation and Forests; Regulation of Torrents, and Arrest of Detritus; Catch-water Drains; Reservoirs for impounding Floods; advantages, objections; Removal of Obstructions; Utility of Movable Weirs; Enlargement of Channel; Straight Cuts; Embankments, low banks, high banks, respective advantages, conditions necessary for high banks, high banks overtopped by rise of flood-level; Instances of Embanking Rivers, Mississippi, Po, Loire. and Theiss; Rising of River-Bed, in Japan, on Yellow River, causes; Pumping; Protection of Upper River Valleys. Ports of Refuge. Concluding Remarks.

FLOODS in rivers are dependent on the amount and continuity of the rainfall over the river basin, the period of the year in which the rainfall occurs, and the nature of the soil and the extent of its previous saturation, as explained in the first chapter. The channel of a river, being formed by the current, is generally only large enough to convey the average discharge of the river, and is quite inadequate to carry off the large floods which occur at intervals in the cold or rainy season. Rivers, accordingly, during these periods, are liable to overflow their banks, and inundate the adjacent low-lying lands. Within the tropics, and in countries exposed to periodical winds, floods are confined to the rainy season; but in the temperate zones, though the great floods, except those of glacier origin, are limited to the cold season, an exceptional coincidence of unfavourable circumstances leads occasionally to the occurrence of a moderate flood in the warm half of the year, especially where the river basin is mainly composed of impermeable strata.

PREDICTION OF FLOODS.

The actual fall of rain, in a given period, over any drainage area furnishes by itself a somewhat incomplete indication of the probable rise of the river draining the district, owing to the uncertainty which exists as to the correct ratio between the actual, and available rainfall in any special basin at any particular period, in consequence of its variation with the season of the year, the previous condition of the soil, and the precise distribution of the rain. In the neighbourhood of the sources of a river, however, the rainfall is the only available index of the advent, and extent of a flood ; whilst the interval between the receipt of the observations of rainfall and the arrival of the resulting flood, is necessarily small, especially near the hilly districts which generally border the limits of a basin, where the valleys are steep and the strata impermeable   Fortunately in these districts the area of land subject to inundation is generally restricted, and its agricultural value comparatively small.

Lower down in the valley, where rich, extensive alluvial plains, formed by the deposits of centuries, border the river on either bank, other means are happily available for giving timely warning of an approaching flood, in addition to rainfall observations. A flood in the lower part of a river is the combined effect of the floods of the upper portion of the main river and the several tributaries, and does not attain its maximum till a considerable time after the maxima have been reached in the branches in the upper part of the basin, the period depending upon the distance to be traversed and the rate of propagation of the flood. The floods, however, of the various tributaries do not generally coincide in the time of their arrival at any given point on the main river, owing to variations in the time of the occurrence of the

flood, partly due to an absence of coincidence in the period
of the rainfall over a large basin, and partly to the difference
in the nature of the strata, and also owing to the differences
in the distances to be traversed, and in the rates of propagation of the floods, which depend upon the fall of each tributary
and the condition of its channel. The floods, accordingly,
from some of the tributaries are generally passing off when
the floods of others come down; and consequently rivers,
in the lower portion of their course, are not usually exposed
to such sudden and proportionately high floods as their
torrential tributaries. Where the distances to be traversed
are not very dissimilar, the floods from the torrential tributaries arrive first, owing to the rapid flow of the rainfall off
the sloping impermeable strata, and the considerable fall
of such rivers; whilst the floods of the tributaries draining
permeable strata descend later, owing to their more gentle
rise and slower rate of propagation. Rivers therefore draining
basins composed of partly permeable and partly impermeable
strata, like the Seine, are less exposed to very high floods
than rivers whose basins are mainly impermeable, such as
the Loire, owing to 'the less liability of coincidence in the
floods of the several tributaries. The actual height, moreover,
attained by a flood depends upon the height of the river
previous to the arrival of the flood, so that the highest floods
generally occur when a second flood comes down before
a previous one has passed off. This is well illustrated by
the flood diagrams of the Seine basin of Nov.-Jan., 1882-83
(Plate 1, Figs. 7-16), where the second flood, though smaller
than the earlier one, attained in most places a greater height.
This circumstance, combined with the influence of evaporation
and the generally low state of the springs during the warm
season, renders high floods extremely rare between May and
October in basins situated like the Seine, containing some
extent of permeable strata (Plate 1, Figs. 3-6); whereas
a winter devoid of floods, such as 1881-82 in the Seine basin,

is quite exceptional. Where, however, the break-up of the winter is somewhat later, as in the upper part of the Mississippi valley, floods sometimes extend into June. A deficiency of rain in the winter, followed by dry weather in the early spring, affords a prospect of a low river towards the close of the autumn, whatever the intervening weather may be, owing to the lowering of the springs by the deficiency in rainfall, which a subsequent increase in rainfall, the effects of which are reduced by active evaporation, is unable to compensate for fully.

By erecting gauges at suitable places on each tributary of a large river, noting the heights attained by floods, and telegraphing the results to stations on the main river, and by observing the periods occupied by the floods on the various tributaries in reaching the several points on the main river, and the corresponding rises at these places in a number of instances, it is possible after a time to predict with considerable accuracy the time of the arrival of the top of the flood at a given place on the main river, and the height likely to be reached by the river some two, three, or more days before the advent of the flood. This system of prediction of floods was established in the Seine basin by Mr. Belgrand in 1854, and has also been adopted in the Loire, the Garonne, and the Saône basins, and arranged for the Ohio ; whilst the floods of the Elbe are predicted at Tetschen and Dresden. It has been brought to such a state of perfection on the Seine, that during the great flood of March, 1876 (Plate 4, Fig. 2), the maximum height of the flood at Paris was predicted, three days beforehand, to within half an inch [1]. On the same occasion, the flood, though greater than any since 1807, did little injury below Paris, owing to the timely warning given of its approach. Thus, in one place an embankment was raised above the predicted flood-level ; and the inhabitants

[1] 'La Grande Crue de la Seine en Mars, 1876.' E. Belgrand and G. Lemoine. Annales des Ponts et Chaussées, 1877 (1), p. 435.

of the low districts had time to remove with their cattle and goods to places of safety. The great value of predictions of the time of the arrival, and probable height of floods when inundations are threatened, is manifest ; and where practicable, this system of warnings should be established on every large river which is liable in flood-time to inundate considerable tracts of cultivated land, and still more for towns and villages which are exposed to danger from floods.

Observations of the rises of the upper portion of a river and its tributaries are also important, in order that the weir-keepers lower down may receive timely notice of the approach of a flood, and thus prepare for its passage by adequately opening their weirs beforehand, a warning specially needed in the summer months when the weirs are naturally closed to maintain the water-level.

PROTECTION FROM INUNDATIONS.

Rivers in their natural condition generally overflow their banks when floods come down, owing to the size of their channels being approximately proportionate to their average discharge ; and, moreover, the material they carry down tends to deposit in the river-bed on the decrease of a flood, thus further reducing the capacity of the channel till scoured away by the next flood. The extent of the inundation depends upon the amount of low-lying land bordering the river, and the height attained by the flood ; it is generally small in the hilly parts of a river basin, owing to the restricted width of the valleys ; but it becomes large in the flat, fertile, alluvial plains stretching out at the base of the hills, where the torrents rushing down from the mountains have their fall abruptly reduced. Floods are not generally injurious when spreading quietly over the land in the cold season, when vegetation is dormant, and where the crops are of a nature to be benefited by the layer of mud which the flood deposits on the land   Some kinds of crops, however, cannot stand a temporary inundation, and most crops are ruined by a summer flood ; whilst some lands lie so low that they cannot be cultivated without protection from floods ; and villages, and even towns sometimes spread over areas within the natural limits of inundation. Works therefore for protection from floods have been often undertaken by the proprietors of the lands bordering rivers, or by the State for the public benefit.

Protection from inundations may be sought, (1) by measures for directly reducing the volume of flood water passing down a river; (2) by facilitating the discharge by the improvement of the channel ; or (3) by works for excluding the river from the low-lying lands. The first object may be attained by encouraging the growth of vegetation, and planting forests on steep mountain slopes, and thus retarding the descent of the flood waters, or by the formation of catch-water drains, or by the construction of impounding reservoirs. The discharge of a river may be facilitated by the removal of obstructions and shoals, by increasing the waterway at weirs and under bridges, and by enlarging the channel. Lastly, the river may be shut off by embankments raised above the highest flood-level, and the drainage of the protected district effected artificially by pumping.

Extension of Vegetation and Forests on Mountain Slopes. In several mountainous regions, vegetation has been reduced by excessive pasturage, and forests have been cut down, producing a denudation of the steep slopes, and a consequent increase in the floods along the valley. Exactly the opposite course should be followed to diminish inundations ; for, as pointed out in the first chapter, the extension of vegetation, and the growth of trees equalize the flow of rivers, by retarding the flow of the rainfall into the watercourses. The general cultivation of districts has similar effects, for the rain sinks more readily into cultivated land ; but the drainage of swampy plains has an opposite influence, for it destroys natural reservoirs of rainfall, and leads the rainfall more or less directly into the nearest stream to swell the volume ·of the floods. The influence of increased vegetation is necessarily slow in manifesting itself, but its operation is sure and permanent; and encouragement has been afforded in some districts, by special legislation, to this means of gradually reducing inundations.

Regulation of Torrents, and Arrest of Detritus. The rapid descent, and erosive influence of torrents may be checked by erecting barriers across their channels. These small cross dams retain the water to some extent, and therefore reduce the velocity of flow and the erosion of the bed. Protective works also against landslips diminish the supply of detritus. In special cases, embankments across the narrow valley of a torrent provide kinds of reservoirs for the storage of detritus; but this system has to be extended from time to time, by erecting additional embankments to provide fresh sites for the accumulation of detritus, when the spaces originally provided have become filled up.

Catch-water Drains. As the discharging capacity of any channel is proportionate to its fall, it is very important to take advantage of any available fall in river valleys where the general fall is extremely small, as, for instance, in the Fen districts of England. This extensive tract of very low-lying land, formerly simply marshes, portions of which were gradually reclaimed from the sea, has been drained by numerous channels leading the water into the enlarged and artificially straightened rivers which flow into the Wash, aided by subsidiary straight drains. Even these straightened rivers have a fall of barely four inches in a mile, and have had to be enlarged in section, and supplemented by drains, in order to convey the drainage waters to the sea. Some portions, however, of the district are at a higher level; but formerly the available fall, due to the higher elevation of these portions, was lost by letting their rainfall flow directly down into the main river, which, by its more rapid descent, rendered the drainage of the lower lands more difficult than if the land had been uniformly flat. By forming catch-water drains, however, contouring the slopes, which collected the rainfall of the higher ground before it reached the flat plains bordering the river, the fall could be utilized; and a smaller channel sufficed for conveying this water to a lower point of the main river which, together with the plains, was relievéd higher up from the influx of this water. Catch-water drains, accordingly, are very useful in providing for the drainage of the higher lands of a flat district in an economical manner, and in relieving the adjacent low-lying plains, and the river itself, from the direct descent of the rainfall of the high lands, which aggravated the danger of inundations.

Reservoirs for impounding Floods. It has frequently been proposed that reservoirs should be formed in river valleys for retaining the excess of flood waters, which the river channel is too small to discharge, and thus prevent their flooding the riparian lands, till the subsidence of the flood enables the impounded water to be quietly discharged without inundating the land. Masonry dams were erected at Pinay and la Roche in the valley of the Upper Loire, early in the eighteenth century, restricting the river to a narrow rocky channel, and thus checking the passage of floods, which appear to have been useful in keeping down the height of large floods in the valley below them. The very high masonry dam across the valley of the river Furens, a torrential tributary of the Loire, at the Gouffre d'Enfer, was designed with the object of preserving the town of St. Étienne from inundation[1]; but previously to its completion, it was determined to utilize the reservoir thus formed for supplying St. Étienne with water, for which purpose a second reservoir has also been constructed higher up the valley. The establishment of

---

[1] Annales des Ponts et Chaussées, 1866 (2), p. 184.

a regular system of reservoirs along the valleys of the
Loire and of the Yonne has been suggested for the purpose
of mitigating the floods in these rivers. Both the Loire
and the Yonne, however, are torrential rivers flowing over
impermeable strata, and with a rapid fall in their upper
portions. It might be possible to find sites in the valleys
of these rivers, or their tributaries, where short high dams
might retain a large volume of water; but unless the water
thus stored up could be utilized, it is doubtful whether,
even on this class of river, the results would justify the
expenditure.

In most cases, the establishment of reservoirs for the
control of floods is open to serious objections. Many river
valleys, as for instance those of the Thames, the Seine, and
the Mississippi, are not suitable for the formation of storage
reservoirs; and in such cases, the cost of making adequate
reservoirs would be far greater than the value of any benefits
they could confer. Also, supposing proper reservoirs to have
been formed, they would be unable to cope with a rapid
succession of floods, for when once filled they would be
unavailable for the succeeding floods. Reservoirs, indeed,
are not economically applicable for dealing with floods,
except under specially favourable local conditions, and unless
they can be also employed for supplying irrigation canals,
or for the supply of water to towns, or as a source of water-
power for electric lighting, or other industrial purposes.
Natural reservoirs may occasionally be obtained, where
barren tracts of low-lying land are capable of improvement
by being flooded by a river and receiving a deposit of
alluvium. The tendency, however, of agricultural improvements lies generally in an opposite direction, namely, the
reclamation of marshy districts formerly serving as natural
reservoirs, which has led to a rise in the level of the floods
in the adjacent valleys, and has increased the damage they
effect. These natural reservoirs form valuable outlets for the

flood waters, and should be retained for this purpose, unless
the increased value of the reclaimed land can be charged
with the requisite compensating enlargement of the channel.

The system of reservoirs for the storage of floods, if it
could be extensively applied, would have a similar regulating
influence on the flow of a river as a large lake; but considerations of expense preclude the due application of the
system, especially in the basins of gently-flowing rivers,
where the extent of land required for reservoirs would bear
no inconsiderable proportion to the lands to be preserved
from inundation. Reservoirs, however, might be established
with advantage in many mountain valleys, where they would
fulfil the double duty of affording an abundant wholesome
water-supply to the nearest towns, and regulating the flow
of a torrent, thus exercising some influence in mitigating
the floods of the main river.

Removal of Obstructions. Any natural or artificial restriction of the channel of a river, produced by rocky shoals,
trunks of trees, the piers of a bridge, or other obstructions,
checks the flow, and consequently raises the water-level and
intensifies the height of the floods above the obstruction.
Each obstruction produces a loss of fall, and therefore a
reduction of the discharge in the river above; for the river
has to rise above the obstruction till a sufficient fall is
obtained at the place to enable the discharge to take place
through the diminished waterway. The removal of these
obstructions distributes the fall in the channel, and consequently increases the discharge, resulting in a lowering of
the water-level and a diminution in the height of the floods.
If, however, the obstructions are only removed from one
portion of a river, the improved discharge in that part will
aggravate the floods in the portion below; and therefore,
in all such works, a river should be dealt with as a whole,
and should consequently be under a single jurisdiction from
its source to its outlet, or at any rate to its tidal limit. The

improvement in the discharge of a river that can be obtained by the removal of rocky shoals, was forcibly shown by the results of the Severn improvement works in 1842, referred to on page 110, where the height of the floods was reduced by the lowering of numerous rocky shoals, notwithstanding the simultaneous erection of a series of solid weirs for the purposes of navigation.

The removal of obstructions from the channel of a river is one of the primary duties of river conservancy, for a river cannot efficiently drain a district if the maintenance of its channel is neglected; and care has to be taken to prevent riparian landowners from restricting the channel for their private interests, to the detriment of other parts of the river, as for instance by the erection of water-mills with inadequate discharge outlets, or by placing fish-traps in the channel, the wire meshes of which soon become choked by weeds, leaves, and other débris, or by encroaching on the river with jetties, quays, or bridges with wide piers and small openings. In mining districts, the refuse is not unfrequently tipped into, or close to the streams, to be carried down by the floods, in reckless disregard of the maintenance of the river below and its outlet, and of the agricultural and commercial interests involved. Every river, with its tributaries, should be protected against such encroachments and injuries, by powers granted to the conservators of the river representing the agricultural and navigation interests of the locality.

Utility of Movable Weirs. The objects aimed at in improving a river for navigation appear somewhat at variance with those for mitigating floods, for the water-level of a river is raised for navigation by means of weirs, and provision is made against an insufficiency of water; whereas, for the mitigation of floods, the discharge is facilitated, and the water-level lowered as far as practicable. Improvements, however, for navigation generally comprise the lowering of shoals, which is also advantageous in promoting the discharge of floods;

and the unfavourable influence on floods of raising the water-level by weirs, can be practically obviated by the adoption of movable weirs, coupled with warnings of the approach of a flood, enabling the flood on its arrival to pass down an almost unrestricted waterway by a timely opening of the weirs. This compensating influence of movable weirs, renders them the equitable accompaniment of improvements for navigation, in order that the prior rights of drainage may be duly safeguarded. Movable weirs, accordingly, should be strictly regarded as proper preventive measures against obstruction of the discharge of a river by improvements for navigation, rather than as direct works for protection from inundations. Nevertheless, navigation works have been so often carried out without due regard to the interests of efficient drainage, that the enlargement of draw-door weirs, and the introduction of movable weirs, have often been subsequently undertaken to mitigate floods. The great value of movable weirs consists in their removing the antagonism between the claims of navigation and drainage, and thus enabling an unrestricted extension of navigation to be carried out without prejudice to the drainage of a district.

Enlargement of River Channel. An obvious, and very efficient method of mitigating floods is the enlargement of the channel of a river, thereby increasing its discharging capacity and lowering the flood-level. This system is only applicable to the prevention of inundations in the summer months, when the floods are generally moderate, and the protection of the crops of most importance ; for the cost of coping in this manner with the larger floods of the cold season would be too great, and inundations in the winter are generally harmless, and sometimes decidedly beneficial. The only objections that can be urged against this method of mitigating floods in the winter, and affording protection to the summer crops, are the cost of maintenance necessitated by a channel enlarged beyond the size maintained by the average discharge, and the probable lowering of the summer low-water level. The system, however, which has been adopted in the Fens, and has been partially carried out on other rivers, is sure in its operation, and can be gradually extended to any desired degree of efficiency.

Discharge increased by Straight Cuts. The adoption of straight cuts in a winding river reduces the length of the channel, and consequently increases its fall and facilitates the discharge. These cuts are most advantageous where the fall is very slight, as for instance on the Fen rivers, where the slightest gain of fall is of importance ; in other circumstances they must be adopted with caution, as the variation they introduce in the flow, and the deviation they produce in the currents, tend to neutralize their good effect. Cut-offs, as they are called in America, have been tried on the Mississippi ; but they are not regarded with much favour, as, though they reduce considerably the length of the channel in the large bends of the river, the labour involved is large, they develop currents unfavourable to navigation, and the changes they produce in the velocity of the flow lead to the formation of shoals below. Moreover, the increase in the discharge is confined to the cut itself, and raises the water-level below it ; whilst the capacity of the channel is somewhat reduced by the diminution in its length. Accordingly, straight cuts are only useful for the mitigation of floods under somewhat exceptional conditions.

Embankment of Rivers. The method of protecting land from floods by erecting embankments along each side of a river possesses several advantages. It provides an enlarged section above for the discharge of the flood waters, and leaves the existing water-levels unaltered. Moreover, by placing the embankments back a certain distance from the banks, the available section of the river can be enlarged to any extent, at the mere cost of the land and of the embankments. Many rivers have been controlled in this manner, as for instance the Po and the Theiss, which, possessing only a small fall in flowing across extensive tracts of level plains, have been systematically embanked. Where low-lying lands border the tidal portion of a river, or where a river flows through flat marshy districts such as the Fens, embankments afford the only available means of preventing inundation. Thus the tidal waters are kept out by banks from the low-lying lands bordering the estuary of the Thames; and the river Witham is embanked from Lincoln to the sea. The Loire also furnishes an instance of a torrential river whose valley is protected from inundations for a considerable length by embankments; but owing to the comparatively small width of the valley, the amount of land protected by the Loire embankments, in proportion to their length, is only one-fifth of the area protected by the embankments of the Po.

Two systems of embanking rivers may be adopted; either low banks may be formed, rising only to a sufficient height to exclude ordinary summer floods, and allowing exceptional summer floods, of rare occurrence, and the high winter floods to pass over; or high banks may be constructed, raised above the highest flood-level, with the object of entirely securing the land from inundation.

The first system has the advantage of involving only a moderate outlay, and of securing the land from inundation at the period when the crops are especially liable to be ruined. Moreover, by allowing the winter floods to pass over the banks, the riparian lands receive each year that deposit of fertilizing alluvium which is beneficial for many crops, and which, if confined to the channel, is liable to reduce its depth on the subsidence of the floods, or to settle at the outlet of the river. Accordingly, low banks along a river are, in most cases, wholly beneficial in providing protection from inundation when most needed, and in not arresting the beneficial distribution of the alluvial matter brought down. Moreover, the landowners being accustomed

M

to inundations in winter, grow crops suited for such conditions; whilst the very exceptional summer floods which may at intervals overtop the banks, involve less loss than the cost of higher banks, and the additional expense of maintenance they would entail. The suitable height of these banks must be deduced from records of the rises of the river in the warm season; and due allowance must be made for the increased rise which the confinement of the river within banks will occasion.

High embankments constructed along a river to secure the surrounding country from inundation, like those erected along the Po, the Loire, the Theiss, and other rivers, have to be made perfectly watertight, and strong enough to resist the water pressure at the highest level; and they are designed to be raised above the limit of the highest possible flood, for if the water once overtops the banks, a breach is soon formed from the washing away of the material by the rush of water down the high bank. Unfortunately this latter condition of security is rarely fully attained, for it by no means suffices to raise the embankments above the highest recorded flood-level. In the first place, floods in a main river result from a combination of the floods of its tributaries; and at rare intervals, an exceptional widespread and long-continued rainfall, following upon a wet season or a heavy fall of snow, together with a coincidence in the arrival of floods from all the tributaries, produces a flood surpassing all previous records. Secondly, the embankments, by confining the river and preventing its flood waters dispersing themselves by flowing over the plains as formerly, materially augment the height of the floods to an extent which is generally underestimated beforehand. Lastly, the detritus brought down by the torrents from the mountains, which previously was distributed over the plains by the floods, is kept within the channel by the embankments, and is liable to settle in the channel as the velocity of the current slackens, in some cases permanently raising the river-bed, and consequently raising the flood-level.

From one or other of these causes, or from a combination of them, these high embankments are sooner or later overtopped ; a breach is formed at a weak spot ; and the liberated flood rushes in an impetuous torrent over the adjacent country, carrying everything before it and causing widespread injury to property, occasiònally accompanied by serious loss of life.

Instances of Embanking Rivers. Longitudinal embankments, or levees as they are termed, have been long resorted to along the Mississippi, for protecting the extensive low-lying alluvial plains through which the river flows between its junction with the Missouri and its outlet in the Gulf of Mexico The most fertile portions of these plains are below the flood-level of the river ; and levees were commenced in 1717 for securing New Orleans, on its establishment, from inundation.   These levees were gradually extended by settlers in the district ; and the total length of main levees now amounts to about 1,300 miles, their prolongation, consolidation, and maintenance having been taken in charge by the Government.   They are essential for the protection of towns built within the area naturally liable to inundation, and they have proved very useful in reclaiming lands liable to be annually submerged ; whilst they have been beneficial for navigation during floods, by marking the line of the channel, and by serving as a quay for local traffic.   The system, however, does not as yet form a perfectly continuous line of protection ; nor have the levees hitherto been sufficiently consolidated throughout to retain the river completely during floods, so that they are always breached in some weak spots when the river is in flood.   Moreover, exceptional floods, occurring on the average at intervals of about ten years, such as those which took place in the spring of 1882 and 1890, are liable to overtop the banks in places, causing the formation of wide breaches, or crevasses as they are termed, in the levees, through which the liberated flood rushes, inundating very large tracts of land.   In 1890 the crevasses in the levees

attained a total length of 6½ miles, a small length compared
to the total length of the levees, but sufficient to set free a very
large volume of water, flooding in one district 1,350 square
miles of country[1].   Owing to the changes which have taken
place in the channel of the river, the levees are in some places
at a distance from the present course of the river.   If it was
desired to secure the land permanently from floods, the levees
would have to be regulated, made continuous, and raised
considerably in places, or placed an ample distance apart, for
in proportion as the gaps are closed, the flood-level rises.
The old levees are being strengthened gradually, and the new
levees are being more substantially constructed ; but it appears
to be considered undesirable to go to the expense of raising
the banks to such an extent as to secure the land from exceptional floods.

The river Po has been embanked for centuries below
Cremona; and the embankments have been gradually extended
towards its outlet, to protect the rich flat alluvial plains of
Lombardy through which it flows on descending from the
Alps.   Its valley, however, has been frequently devastated by
floods, several considerable inundations having occurred during
the nineteenth century.   One of the worst of these took place in
October, 1872, when the river forced its way through thirty-
three breaches in the embankments which had been overtopped ; and the inundated land was not entirely free from
water till the following April.   Conflicting views have been
held as to the raising of the bed of the Po by the deposit of
sediment in its embanked channel ; but the silting up of its
channel has undoubtedly been slight, and the rise in its flood-
level, amounting to about 7 feet in the last two centuries, must
be mainly due to the increased restriction of the flood waters
by the extension of the embankments along the river and its
tributaries, aided by improved drainage and the cutting down

[1] 'Report of the Mississippi River Commission for 1890;' 'Report of the Chief
of Engineers, U. S. A. for 1890,' p. 3083.

of forests. The Commission appointed by the Italian Government, after the flood of 1872, to consider the best means of
preventing the inundations of the Po, considered that it was
unadvisable to endeavour any longer to confine the floods
within the river channel by adding to the height of the
embankments, as such a course tended to aggravate the
injuries which it was designed to prevent. Nevertheless,
the raising of the embankments up to the flood-level of 1872
has since been partially carried out.

Since 1872, large breaches have been formed in the embankments of the Po in 1879, of the Adige and other Venetian rivers
in 1882, and of the Reno in 1889. Moreover, floods rising
higher than the great flood of 1872 have occurred on all these
rivers and their tributaries since that period. Thus the Po
rose slightly higher at Pisa in 1879 than in 1872, 1 foot
5 inches higher in 1886, and 1⅔ feet higher at Milan in 1887;
the Adige rose 4½ feet higher at Verona in 1882 than in 1872,
and the Reno 2¼ feet higher at Ferrara in 1889; whilst some
of the rivers of this part of Italy have exhibited a still greater
increase in height since 1872 [1]. To remedy this perilous
condition, the Italian Government are attacking the evil at
its source by retarding the influx of torrential affluents by
constructing small cross dykes, by endeavouring to reduce
gradually the quantity of detritus brought into the rivers by
planting forests on the denuded slopes of the hills, and by
executing works to prevent landslips. The sum of £1,150,000
was also expended on embankment works in 1888–90, since
which period little has been done owing to financial difficulties;
and out of a total length of embankments on these rivers of
4,012 miles, there are still 223 miles of embankments liable
to be overtopped by the highest floods.

The Loire, below its junction with the Allier, flows through
a plain subject to inundations, which has been protected by

---

[1] I am indebted to Cav. Luigi Luiggi, of the Ministry of Public Works at Rome,
for these particulars about recent floods.

embankments. In most cases, a bank on one side only has been required, as high ground generally borders the opposite side of the river. Great floods, however, occurring about every ten years, come down with a rapid flow owing to the fall of the river, and usually burst the banks and overflow the valley. The remedy proposed consists in forming openings in the banks at suitable places, protected by pitching, with their sills just above the height of ordinary floods, so that the higher floods may flow comparatively quietly through these openings, and in this way the devastation caused by the sudden bursting of the river through breaches may be avoided [1]. The land-owners have, however, hitherto refused to allow outlets to be formed in front of their lands.

The river Theiss, on descending to the plains of Hungary, has a very small fall, amounting to from 6 to 9 inches per mile above Szegedin, and to only 1·2 inches between Szegedin and its junction with the Danube. In consequence, the plains through which the river winds were frequently flooded ; and the extent of country covered by these floods was 6,000 square miles. Of this area, 4,200 square miles have been protected by the regulation and embankment of 740 miles of river. The works of improvement, however, having been arrested in 1867, and the necessary works for counteracting the gradual rise of the flood-level, resulting from the straight cuts formed above the town, having been neglected, the waters of the Theiss rose above the protecting embankments in March, 1879, and, forming a breach, destroyed a portion of the town of Szegedin, and inundated two hundred square miles of country. The raising of the embankments, and the completion of straight cuts below the town have since secured Szegedin from inundation, though the floods of 1881 and 1888 rose higher than the flood of 1879, nearly reaching the top of the embankment on the right bank in some parts below the town.

[1] ' Rivières et Canaux.' P. Guillemain, vol. i, p. 247.

Raising of River-Bed. A point of considerable importance
in considering the expediency of protecting districts from
inundation by longitudinal embankments, and the prospects
of permanent protection afforded by such works, is the possibility of deposits of alluvium in the river-bed gradually raising
the bed, and consequently producing a rise in the flood-level,
which would necessitate successive additions to the height of
the embankments, or would result eventually in the rupture of
the banks by the flood overtopping them. The Rhine has
been systematically embanked for long distances without
any elevation of its low-water level having been observ-
able ; and if any raising of the bed of the Po has occurred,
it has unquestionably been slight, though this possibly
may have been due to the periodical scouring out of the
channel by the escape of the river through breaches in
the banks. On the other hand, the river Reno in northern
Italy raises its bed in the plains with the detritus it
brings down from the mountains, which it deposits when its
velocity is checked by the reduction in its fall, and is a source
of anxiety in consequence.

Some rivers in Japan, descending from the mountains, and
embanked across flat alluvial plains, have so raised their beds
by the gradual deposit of their sediment, followed by a raising
of the protecting embankments, that in some cases the surface
of the water is 40 feet or more above the level of the plains over
which the rivers flow. Under these peculiar circumstances, in
constructing railways across these plains, it has occasionally
proved advisable to drive a tunnel on the level underneath the
bed of the river, rather than raise the railway to cross over the
river[1]. It is evident also from an account by Mr. Morrison,
an English engineer residing at Shanghai, of a personal inspection of the Yellow River of China, in September 1888,
after the great inundation which resulted from a breach in

---

[1] 'Railway Work in Japan.' W. F. Potter. Minutes of Proceedings Institution
C. E., vol. lvi, p. 10, plate 1, figs. 1 and 2.

the embankments of the river in the autumn of 1887, that
a similar raising of the river-bed occurs on this river[1].

The embanking of a river tends rather to improve its scouring
capacity by concentrating its flood waters, thereby generally increasing its hydraulic mean depth and consequently its velocity;
and therefore the embankments somewhat diminish a river's
tendency to deposit its sediment. The limits, however, of deposit
are restricted by the embankments; and the alluvial matter
which, under natural conditions, would spread over the inundated
plains, is confined within the flood channel by the embankments. Accordingly, torrential rivers descending from the
mountains, heavily charged with detritus, when their velocity is
diminished by the reduction in the fall, deposit the heavier
material in their bed which they formerly strewed over the plain,
in spite of a larger proportion of the alluvium being carried
down by the concentrated current. The rising of the beds of the
Japan rivers referred to, and of the Yellow River, is clearly due
to this cause; and under such conditions, the raising of the
embankments to correspond with the rise of the bed, infallibly
leads sooner or later to a breaking of the banks, and the discharge of the flood waters of the river into the plains below,
a catastrophe which periodically occurs in Japan and in the
valley of the Yellow River, spreading devastation far and wide.

Pumping for Drainage of Land. In some places the land
is at such a low level in relation to the watercourses of the
district, that it cannot be drained by gravitation; and the
drainage waters have to be raised over the protecting banks
by pumping, and discharged into the nearest watercourse.
This system is extensively adopted in Holland, and it has
been applied to portions of the Fens in England. Low-lying
land can thus be reclaimed and secured from inundation,
provided its value is sufficient to bear the cost of the pumping,
which is sometimes reduced by the use of windmills to
supplement steam power.

[1] 'North China Herald.'

Protection of Upper Valleys of Rivers. Embankments and regulation works should not generally be undertaken in the upper part of river valleys, for they hasten the descent of floods and promote the carrying down of detritus, and consequently intensify the floods in the lower parts of the valleys, and thus protect the less valuable lands above at the expense of the alluvial plains below. The Rhone valley, indeed, above the Lake of Geneva, has been protected from inundation by regulation works and embankments, without injury to the river below, owing to the intervention of the moderating influences of the Lake of Geneva. This, however, is an exceptional instance, where the peculiar physical conditions rendered such a course practicable without detriment to the lower parts of the valley. The Rhone also is a river along whose valley protection from inundation is specially important, as, owing to its glacier origin, its floods occur in the summer season.

Refuge Ports on Rivers. Where the navigation on large rivers is liable to be endangered during floods by débris and floating ice, ports of refuge are constructed at suitable places near towns, where vessels can take shelter during the winter. A protecting bank or jetty is constructed enclosing a portion of the river alongside one of the banks, having a suitable depth, thus forming a sheltered haven with an entrance at its downstream end, in which vessels can lie in safety. Moreover, by the construction of a quay along the land, the port furnishes a convenient sort of dock for loading and discharging vessels. A port of this kind has been established on the Main, a little below Frankfort (Plate 4, Fig. 5); and many ports of refuge have been constructed on German, and most American rivers.

Concluding Remarks. In matters relating to river conservancy and works for protection against floods, every main river basin should be under the control of a single body, either the State or conservators representing the several districts, who alone can undertake systematic works for the

general benefit of the whole watershed. This controlling
authority should be vested with full powers for preserving
the channel of the river from injury, and to prevent works
being undertaken by private individuals; for refuse thrown
recklessly into a river is injurious to the maintenance of
the channel lower down, and isolated embankments or cuts,
affording protection or relief at one part of the valley, serve
to intensify the floods elsewhere. The management of these
matters, on the continent of Europe and in the United States,
by competent officials under the direction of the State, affords
evidence that the control of rivers may be fairly regarded as
of national importance, to be carried out for the public
benefit. The comprehensive nature of the works involved,
necessitating their being taken out of the hands of private
landowners, obliges public funds to be employed for their
execution, either drawn from the taxes or raised by a general
rate levied on the district. In any case, however, it would be
equitable for the proprietors of lands liable to inundation to
contribute a considerably larger proportion of the cost of
works for the mitigation of floods, than the owners of the
higher lands, who are merely interested in the general
healthiness of the district and the preservation of the means
of communication. Absolute security from inundation should
be provided for towns in the neighbourhood of rivers; but
often, in the case of cultivated land, the cost of works
providing entire immunity from floods would be greater than
the loss resulting from partial inundations during exceptional
floods. The removal of obstructions, and the improvement of
the channel are the first works to be undertaken for the
mitigation of floods. Further protection against inundation
can be provided to any extent by continuous embankments
along the main river and its tributaries, aided sometimes by
straight cuts where the river is tortuous and the fall very
slight. Embankments, however, must be adopted with
caution, for they often have to be raised much higher than

anticipated at the outset, owing to the rise in the flood-level;
and they cannot readily be abandoned when the riparian
landowners have become accustomed to their protection,
even when, as in the case of the Loire, periodical disasters
counterbalance their advantages. Embankments are very
advantageous where wide tracts of low-lying land border
a river, such as the alluvial valley of the Mississippi; but they
are generally inexpedient in the higher parts of a valley,
owing to the increase they produce in the floods below, or in
the somewhat narrow valley of a torrential river which
periodically resumes its sway over the valley. This system
of protection should be avoided, if possible, in the case of
rivers which raise their beds in the plains with the detritus
they bring down from the mountains, for under these circumstances the struggle with nature is continuous; and she,
sooner or later, avenges herself for the opposition to her laws,
by bursting her bonds and inflicting widespread disaster, the
struggle being subsequently renewed, to be terminated in
a similar manner.

Mitigation of floods near the mouth of a river results from
improvement works at the outfall, though generally the object
of such works is solely the benefit of navigation. Regulation works, indeed, at the confluence of two rivers, to prevent
the injurious conflict of their currents, are sometimes undertaken to promote the discharge, as well as for navigation;
and the new cut for the outfall of the Witham was carried out
with this double object. Most of the improvement works,
however, at the mouths of rivers, such as the dredging of the
Tyne, merely incidentally reduce floods by facilitating the
discharge through the enlargement of the waterway. Outlet
works would usually be too costly, and their influence would
not extend sufficiently far up a river, to be undertaken for the
mitigation of floods; but in benefiting navigation, they also
exercise a beneficial influence in moderating floods.

CHAPTER VIII.

DELTAS OF TIDELESS RIVERS; AND IMPROVEMENT OF THEIR OUTLETS.

Contrast between Deltas of Tideless Rivers, and Tidal Estuaries. *Deltas:*—Formation, Form, Bars; Advance; Rhone Delta; Nile Delta; Danube Delta; Volga Delta; Mississippi Delta; Comparison of these Deltas; Methods of Improvement. *Dredging and Harrowing on Bar :*—Objects; at Outlets of Danube and Mississippi; at Volga Mouths. results, proposals. *Parallel Jetties at Outlet :*—To scour the Bar; Conditions of success, and permanence; Rhone Jetties, previous conditions of Mouths. works carried out, their effects, causes of failure; Danube Jetties, Sulina Mouth selected, form of works, results, causes of success, compared with Rhone works, gradual shoaling in front, improvement of Sulina branch; Mississippi Jetties, South Pass Outlet selected, works described, results, works at Head of Pass, shoaling in front of outlet, dredging, prolongation of jetties imminent; Mississippi and Danube works compared. *Ship-Canal instead of a Delta Channel :*—Object, at Ostia, San Carlos Canal, Mahmoudieh Canal, St. Louis Canal, proposed for Danube and Mississippi. Conclusions.

ALL rivers, till they approach the sea into which they finally discharge the rainfall they have brought down, present somewhat similar characteristics; their current always flows in one direction; their channels bear a proportion to the size of their basins; and the variations in them are merely due to differences in climate, rainfall, stratification, and the fall of their bed. The detritus they bring down is gradually carried along to their outlet, or strewn over the adjacent plains; and their channel is more or less uniformly maintained by the average discharge. Rivers, however, on approaching the sea exhibit a notable difference in their condition, according as they flow into a tideless, or a tidal sea. In the first case, the river continues its downward flow with a diminishing fall and velocity, till, on encountering the inert waters of the sea, its current is checked, and finally arrested as it mingles its waters with the ocean. In the second case, the tide flows and ebbs

for some distance up a river, the extent of the tidal influence
depending upon the inclination of the river-bed, the rise of the
tide at the outlet, and the volume of fresh water coming
down the river. Accordingly, in a tideless river, the outlet
channel is wholly maintained by the fresh-water discharge, as
in the portions higher up ; whereas in a tidal river, the water
brought down by the river is largely reinforced by a great
volume of sea-water, which enters the estuary during the
flood tide, and flows out, together with the accumulated
volume of fresh water, during the ebb. These differences
produce a very marked contrast in the form and constitution
of the outlets of sediment-bearing tideless, and tidal rivers ;
and they necessitate the adoption of essentially different
methods for the improvement of these two classes of rivers.
A comparison of the tideless mouths of the Rhone, the
Danube, and the Mississippi (Plate 5, Figs. 1, 2, and 5) with
the estuaries of the Thames, the Humber, the Mersey, the
Clyde, the Seine, the Loire, the Weser, and other tidal rivers
(Plates 7, 8, and 9), fully exemplifies the great contrast which
they present. There are, indeed, as pointed out in Chapter I,
p. 19, some rivers whose outlets occupy a sort of intermediate
position between tideless and tidal rivers, in consequence of
either a great fresh-water flow densely charged with alluvium,
like the Ganges, or a small tidal rise in proportion to the
fresh-water discharge, like the Rhine ; or a tideless river may
be free from a delta, owing to the small proportion of sediment
in relation to the discharge. In dealing, however, with the
improvement of the outlets of rivers, it is expedient to draw
a broad distinction between the two extreme classes of rivers,
and to consider them quite separately, reserving a modification
of the principles of improvement for those river mouths which
hold a somewhat intermediate position.

DELTAS.

A very marked peculiarity of rivers bringing down large

quantities of alluvium, collected in the upper part of their valleys and eroded from their banks, is the deltas which they form at their outlets, when flowing into tideless seas. The detritus brought down by a river partly in suspension and partly rolled along the bottom, at length reaches the sea, and is gradually deposited at the outlet, when the current of the river is arrested on emerging out of a confined channel into the open sea. This deposition is promoted by the much more rapid settlement of clayey alluvium in salt water than in fresh, owing probably to the aggregation of the particles produced by saline solutions. The denser material is first deposited; whilst the lighter matter is carried on further by the fresh water which spreads out like a fan on the top of the salt water. but is eventually deposited, if not carried away by any littoral current. Accordingly, the continually accumulating mound of sediment advances in a fan-shaped form into the sea, projecting out from the coast-line, and reduces the fall of the river near its mouth by prolonging the outlet channel. The enfeebled current finds its way to the sea, through the mass of sediment which it deposits, in several shallow, diverging channels, which, with the coast-line, present some resemblance to the Greek letter Δ ; and, in consequence, the name delta has been given to the flat, low-lying alluvial land formed by these rivers. The discharge through each of these channels, in depositing its burden of sediment on emerging into the sea, forms a bar with the denser material in advance of the outlet; so that these tideless rivers offer serious obstacles to navigation at their mouths, though often possessing an excellent navigable channel above their deltas. These bars progress seawards with the advance of the delta; and their distance from the mouth of the channel depends upon the volume and velocity of discharge through the channel. The rate of advance of a delta depends upon the amount of sediment brought down by the river and its density, the depth of the sea in front, and the influence of disturbing causes, such as winds, waves, and littoral currents. The advance of the delta at each mouth of a tideless river is proportionate to the discharge through the mouth, for the sediment conveyed through each outlet channel of any particular river is proportionate to the discharge.

Delta of the Rhone. The Rhone divides into two branches a little above Arles, at which point its delta is considered to commence by the divergence of the Little Rhone from the main river. The Great Rhone flows past Arles in a channel 500 feet wide, and having a maximum depth of 46 feet; and, though Arles in Roman times owed its importance to its proximity and ready access to the sea, it is now 32 miles distant by river from the mouth of the Great Rhone, owing to the steady advance of the delta. Previously to 1852, the Great Rhone, which is the navigable branch, and conveys over four-fifths of the discharge of the river, flowed into the Mediterranean Sea, at the Gulf of Lyons, through six mouths; and the advance of its delta has been estimated at 140 feet annually (Plate 5, Fig. 1). The Rhone brings down large quantities of mud, sand, and shingle; but the shingle finally disappears at Soujean, $4\frac{1}{2}$ miles above Arles; and the sediment below this point consists entirely of mud and sand, of which the whole delta is composed. The main branch of the Rhone brings down about $23\frac{1}{2}$ million cubic yards of alluvium annually, as estimated by Mr. Guérard, or about $\frac{1}{2166}$ of the volume of water discharged; and this alluvium for the most part consists of fine sand rolled along the bed of the river, less than a fourth of the volume of silt brought down having been found in suspension in the current[1].

The sea-bottom along the Mediterranean coast in front of the delta has a fair dip seawards, a depth of 5 fathoms being reached within 5 furlongs from the shore; but the slope is considerably flatter in the inner bay, called the Gulf of Foz,

---

[1] Rapport sur l'amélioration des embouchures du Rhône. A. Guérard, 1888; and Minutes of Proceedings Institution C. E., vol. lxxxii, p. 309.

into which the existing outlet channel discharges. The depth over the bar in front of the principal mouth, in 1852, previously to the commencement of any improvement works, averaged 4 feet 7 inches, increasing to a maximum of $7\frac{1}{3}$ feet in flood-time, and falling to a minimum of 1 foot 7 inches during the low stage of the river. Access from the river to the sea being thus practically barred, the principal port of France has been established on the sea-coast at Marseilles, out of the reach of the alluvium of the Rhone, where the necessary shelter has been provided artificially by large breakwaters, which rivers commonly afford on tidal coasts.

Delta of the Nile. The Nile furnishes one of the best known instances of a delta-forming river; and to its outlets the Greeks applied the name of delta, which has come into general use to denote similar excrescences of other tideless rivers. The river begins to divide at Cairo; and about 12 miles lower down it splits into two principal branches, called Rosetta and Damietta, through which, in conjunction with several minor channels, it finds an outlet for its waters into the Mediterranean, some ancient branches and mouths having silted up. The delta of the Nile extends over an area of about 9,000 square miles, having been gradually formed by the alluvium of the river, which during high Nile constitutes about $\frac{1}{660}$ of the discharge.

The annual rise of the Nile is very regular, being caused by the tropical spring rains over the mountains of Abyssinia, from whence the alluvium is derived. The river begins to rise about the beginning of July; it reaches its maximum height in September, averaging about 25 feet at Cairo; and it falls again more gradually to its minimum flow in April. Its average maximum discharge at Cairo is about 11,000 cubic yards per second, and its minimum, 525 cubic yards. Accordingly, the flood discharge is twenty-one times the low-water flow, and raises the river 3 or 4 feet at its mouths; so that whilst there is a depth of about 7 feet over the bars at

the outlets when the flood is at its maximum, the entrance
is barred, even to small vessels, during low Nile.

Delta of the Danube.   The Danube flows past Isaktcha in
a single channel, 1,700 feet wide and 50 feet deep; but 15
miles lower down, its delta commences by the splitting up
of the river into two channels, about 45 miles from the Black
Sea in a direct line (Plate 5, Fig. 2).   The southernmost
channel divides into two branches 11 miles lower down, so
that eventually the river flows through its delta in three main
branches to the sea.   The northern or Kilia branch, carrying
nearly two-thirds of the whole discharge of the river, after
meandering in places through several channels, forms an
independent delta near its outlet; and, after a course of
61 miles, it flows into the Black Sea through twelve mouths,
encumbered by a continuous shoal of alluvium in front of
their outlets, with depths over it of only from 1 foot to 6 feet,
and advancing seawards 200 to 300 feet annually[1].   The
central Sulina branch, receiving rather less than one-thirteenth
of the discharge of the river, flows into the sea through
a single outlet, 63 miles from the head of the delta, having
a bar in front, over which the depth varied from 7 to 12 feet
between 1829 and 1857, the bar attaining its greatest height
on the subsidence of high floods.   The yearly advance of the
delta in front of this mouth averaged 94 feet before the
commencement of the improvement works; this much smaller
rate of advance, as compared with that of the Kilia delta,
being due to the much smaller discharge of the Sulina channel,
and the greater depth of the sea in front of its outlet.   The
southern, St. George's branch carries rather more than two-
sevenths of the discharge, and possesses a good navigable
channel with a minimum depth of 16 feet; but it splits up
near its outlet into two branches, and discharges into the

---

[1] 'Description of the Delta of the Danube.' C. A. Hartley. Minutes of
Proceedings Institution C. E., 1862, vol. xxi, p. 278.

Black Sea, 75 miles from the head of the delta, through two
mouths, having depths of only 3 feet and 7 feet respectively
over their bars, which are further out from the shore than the
Sulina bar.

The area of the delta of the Danube comprised between the
northern and southern branches is 1,000 square miles. The
total volume of water discharged by the Danube averages
4,630 cubic yards per second at low water, and 12,000 cubic
yards at high water; but at an extremely low stage, the
discharge falls to nearly half the first discharge, and in an
extraordinary flood it reaches over three times the latter
discharge, whilst the mean discharge is 8,240 cubic yards per
second. The alluvium carried in suspension by the river is
naturally very small during the low stage of the river, and
very large proportionately during floods. The suspended
matter amounts, on the average, to $\frac{1}{8100}$ of the volume of
water discharged; but a considerable amount of denser
material must be rolled along the bed during floods, which
materially increases the proportion of alluvium actually
brought down.

Delta of the Volga. The delta of the Volga commences
about 31 miles above Astrakhan, where the Buzan branch
diverges from the main channel; and its area amounts to
about 5,300 square miles. Near Astrakhan and below, other
channels branch off, which again subdivide, so that eventually
the Volga enters the Caspian Sea through at least two hundred
mouths[1]. The Volga has an unusually moderate fall, even
in the upper portion of its course; and the large quantity of
alluvium it brings down appears to be in great measure due
to the erosion and falling-in of high cliffs which border its
channel in some parts. Uncertainty exists as to the proportion of alluvium brought down; but, according to some

---

[1] 'Les Embouchures du Volga.' V. E. de Timonoff. Vme Congrès International de Navigation intérieure, Paris, 1892.

observations, the river has carried along 26,000 cubic yards per day during a flood, which bears a somewhat small proportion to the maximum discharge of the river in flood-time, averaging about 27,000 cubic yards per second. At the low-water stage, the discharge falls to 2,600 cubic yards per second. The rate of advance of the delta of the Volga, which necessarily varies at the different mouths, is not accurately known; but it has been estimated at 1,270 feet per annum, a very rapid advance, which, if not considerably exaggerated, can only be accounted for by the shallowness of the Caspian Sea in front of the delta. The northern portion of the Caspian Sea is, indeed, very shallow, and there has been an average annual decrease in its depth of 3 to 9 inches in the last fifty years; and the level of the sea is affected to the extent of some feet by the wind. Out of the numerous changeable branches of the Volga traversing the delta, two only at the present time serve for the navigation between Astrakhan and the sea, namely, the Bakhtemir and Kamysiak channels. The former channel is the most frequented, as a normal depth of 8 feet over its bar has been obtained by dredging, whereas the depth over the Kamysiak bar is only 4½ feet; and these depths are liable to be raised or lowered two or three feet by the wind. In other respects, however, the Kamysiak is the better channel, for it is much straighter and shorter, and generally deeper than the other; and sea depths of 3 fathoms approach much nearer the Kamysiak mouth than the other outlets. The advance also of the delta must be much more rapid in front of the Bakhtemir outlet, as rather over one-third of the whole discharge of the Volga is conveyed by this channel.

Delta of the Mississippi. The average rise of tide in the Gulf of Mexico opposite the mouths of the Mississippi is only 14 inches, and there is only one tide in a day. Accordingly, the Mississippi entering a practically tideless sea, and bringing down large volumes of sediment during floods, forms a delta like the European rivers previously described, which flow into tideless inland seas. The delta of the Mississippi is considered to commence just below the confluence of the Red River, 316 miles from the outlet, where the first small branch separates from the main river; and it stretches over an area of 12,300 square miles[1]. The regular splitting up, however, of the main channel does not take place for a long distance further down; and at Fort St. Philip, within 35 miles of the Gulf of Mexico, the river still flows in a fairly uniform channel, 2,470 feet wide and 120 feet deep (Plate 5, Fig. 5). Twenty miles below Fort St. Philip, the river, having attained a width of about $1\frac{3}{4}$ miles, and its depth being reduced to 30 feet, divides into three very diverging channels, two of which eventually split up again, so that at last it discharges into the Gulf of Mexico through seven mouths. The delta projects about 200 miles in front of the natural shore-line of the gulf; and its formation has extended over at least 4,400 years. The mean discharge of the Mississippi into this main delta amounts to about 23,000 cubic yards per second; and the average quantity of alluvium discharged into the Gulf of Mexico in a year reaches $299\frac{1}{2}$ million cubic yards, giving a proportion of sediment equivalent to about $\frac{1}{2420}$ of the discharge. The sediment consists of silt, sand, and clay; and the heavier materials are rolled along the bed of the river, the quantity thus transported being estimated at one-tenth of the whole. The annual advance of the delta at the three principal outlets is, 300 feet at the mouth of the South-West Pass, which conveys one-third of the whole discharge; 260 feet at the mouth of the Pass à l'Outre, which carries one-fourth of the discharge; and it was 100 feet at the mouth of the South Pass before any works were commenced, only about one-tenth of the discharge passing through this outlet, the advance being

---

[1] 'Report on the Physics and Hydraulics of the Mississippi River.' Humphreys and Abbot, p. 452.

naturally somewhat proportionate to the discharge. Though
the average depth in the passes is from 34 to 58 feet, the
bars in front of the outlets reduced the available depth to
a maximum of about 13 feet, which is the depth over the bar
of the South-West Pass about 5 miles seawards of the outlet,
constituting formerly the only navigable channel between the
river and the gulf, after the deterioration of the outlet of the
North-East Pass, one of the branches from the Pass à l'Outre,
during the first half of the nineteenth century.   Owing to the
smaller discharge through the South Pass, the depth over its
bar was only 8 feet; but the bar was only $2\frac{1}{4}$ miles in advance
of the coast-line, less than half the distance seaward of the
bar of the South-West Pass.   The sea-slope in front of the
delta of the Mississippi is about the same as that of the
Mediterranean Sea in front of the Rhone delta, and about
four times as steep as the slope of the Black Sea in front of
the Danube delta.

Comparison of Deltas. The physical conditions which it
is important to compare in the foregoing examples of deltas,
are the proportionate volume and density of the sediment,
the inclination of the sea-slope in front, the rate of advance
of the delta, the natural depth over the bar, and the extent
and direction of any littoral current across the face of the
delta.

The Rhone appears to carry along about one-eighth more
sediment, in proportion to its discharge, than the Mississippi,
and probably between two and three times more sediment
than the Danube. This sediment, moreover, seems to be
considerably denser in the case of the Rhone than that of the
Mississippi; whilst the alluvium of the Danube is, for the
most part, much lighter. Accordingly, the condition of the
Rhone as regards sediment is more unfavourable, both as
to quantity and density, than that of the Mississippi; whilst
the condition of the Danube is superior in both these respects
to the Rhone and the Mississippi. The flatness, however, of

the sea-slope in front of the Danube delta places the Danube in this particular at a disadvantage in comparison with the Rhone and the Mississippi; but the Volga is considerably more unfavourably affected by the shallowness of the northern part of the Caspian Sea. The Volga, on the other hand, appears to carry down a comparatively small proportion of sediment; whilst the Nile seems to be more heavily charged than the other rivers described. The advance of a delta depends on the volume of sediment deposited, and the sea-slope; and thus the rates of advance of the Mississippi and Danube deltas are fairly similar, owing to the different proportions of sediment being compensated by differences of sea-slopes. Though the proportion of sediment in the Volga appears to be considerably less than even in the Danube, the advance of its delta is probably more rapid, owing to the flatness of the sea-slope in front of its mouths. The advance of the several branches of a delta varies with the relative discharges; and though a larger discharge provides generally a better depth over the bar, it is accompanied with a more rapid advance of the mound of deposit, and with a greater protrusion of the bar out to sea.

A littoral current retards the advance of a delta by sweeping away the lighter alluvial materials brought within its influence. An easterly littoral current carries the turbid waters from the Damietta mouth of the Nile across the entrance to Port Said harbour (Plate 13, Fig. 4); a southerly current exists in front of the Danube delta, which is aided by the prevalent north winds; and a permanent current, extending to depths of about 330 feet, travels from east to west along the Mediterranean shore of the Rhone delta, which may account for the smaller rate of advance of this delta in former times as compared with the average advance of the Mississippi delta, in spite of similarities as regards proportionate sediment and sea-slope. There appears to be no natural littoral current along the northern coast of the Caspian Sea; and the prevalent winds, blowing up and down the Volga delta, cannot create one. A westerly current flows across the outlets of the Mississippi, mainly produced by the prevalent winds, aided somewhat by the feeble tidal wave entering the Gulf.

Methods of Improvement. The above descriptions of deltas show what serious obstacles to navigation these alluvial accumulations, formed by the rivers themselves, present. These natural deposits cannot be prevented; and their gradual prolongation increases the difficulties of the situation by lengthening the channels, and reducing by degrees the fall of the river near its outlet, promoting the subdivision and shoaling of the branches. The largest rivers possessing an excellent navigable channel for great distances inland, may thus be practically cut off from proper access to the sea, and thereby debarred from external communications. The removal, therefore, of these barriers to ocean-going trade is of the highest importance to the extensive districts served by such rivers.

Three methods have been adopted for securing improved navigable communication between these rivers and the sea, namely: (1) Dredging or harrowing on the bar; (2) Prolongation of outlet channel by parallel jetties; and (3) Construction of a ship-canal. The first method aims at the direct lowering of the obstruction at the bar by dredging, or by stirring up the material so that it may be carried away by the outgoing current. The second system effects the increase in depth by the scour of the concentrated current across the bar. By the last plan, the difficulties of the delta are evaded by connecting the river, above the delta, with the sea by a ship-canal emerging at a point of the coast beyond the influence of the alluvial deposits. The deepening of the channel over the bar of one of the outlets was, in the first instance, attempted, chiefly by harrowing, at the Danube and Mississippi deltas; and dredging has been adopted for improving one of the main outlets of the Volga delta.

Embankments were constructed for prolonging one of the outlet channels of the Rhone, and thus directing the issuing current across the bar ; and jetty works have been carried out with the same object at the mouth of one of the branches of the Danube, and of the Mississippi. Lastly, a ship-canal was proposed for affording an outlet for the trade of the Danube, and also of the Mississippi, beyond the zone of deposit of the delta ; and ship-canals have been made for connecting the Rhone and the Nile with the sea beyond the limits of their deltas.

DREDGING AND HARROWING ON THE BAR.

The direct removal of the crest of the bar in front of the outlet of one of the channels of a delta, to a depth suitable for navigation, by dredging or harrowing, was naturally the expedient first resorted to.

Dredging and Harrowing at Outlets of Danube and Mississippi. The Sulina branch of the Danube was the only navigable channel to the sea in 1857 ; and as a first attempt at improvement, a rake was dragged across the bar, according to the old Turkish method for providing a temporary deepening ; and a dredger was also employed for throwing up and putting into suspension the material forming the bar. These measures, however, produced only a transitory improvement of the outlet channel ; for the river current, becoming enfeebled on emerging into the sea, was unable to retain in motion all the material it had brought down, and therefore was incompetent to bear an increased burden into deep water. A similar system was resorted to at the bar of the South-West Pass of the Mississippi delta, on a larger scale, between 1852 and 1875, harrowing on the bars having been tried as early as 1726 ; but an increase in depth, from 13 to 18 feet, was with difficulty obtained, for a width of 300 feet, by large dredging operations ; and the channel soon reverted to its original average depth of 13 feet whenever the dredging was suspended.

Dredging at the Mouths of the Volga. The depth over some banks obstructing the mouth of the Bakhtemir branch of the Volga delta was formerly little over 4 feet at the normal water-level, and was reduced to 2 feet by land winds. These banks have been lowered by dredging, since 1874, so as to provide a depth of 8 feet at mean level, for a width of 420 feet, the total length of dredged channel over the several banks amounting to 5⅜ miles; and the maintenance of this depth is effected by raising 76,000 cubic yards annually with one dredger[1]. To obtain, however, an additional foot in depth in the outlet channel of this branch, it would be necessary to dredge along a length of at least 17 miles. Moreover, the advance of the delta in front of this branch must be more rapid than at other parts, as one-third of the total discharge of the river flows through this outlet ; the branch is narrow and winding in places ; there are some shoals in it which have to be dredged; and the distance by this route between Astrakhan and the sea is about 67 miles. Accordingly, it is now proposed to revert to the smaller Kamysiak branch, at one of the mouths of which a futile attempt was made, in 1858–69, to deepen the channel over the bar by directing the current between training banks of fascine mattresses aided by dredging, and concentrating a larger discharge in the selected channel by closing the minor outlets.

The Kamysiak branch has a minimum depth of 14 feet between Astrakhan and its bar ; its channel is fairly straight; and its bar, with a normal depth over it of 4½ feet, is about 1½ miles wide, and therefore considerably narrower than the banks which impeded the mouth of the Bakhtemir branch. Moreover, the 3-fathom line is much nearer this outlet; and the distance from the sea to Astrakhan by this channel is only about 46 miles. Owing also to its much smaller discharge than that of the Bakhtemir branch, the advance

---

[1] 'Les Embouchures du Volga.' V. E. de Timonoff. Vme Congrès International de Navigation intérieure, Paris, 1892.

of the delta at its outlet must be much slower. It is, accordingly, believed that a 14-foot channel might be obtained at one of the mouths of the Kamysiak branch by adequate dredging plant, which would suffice for vessels navigating the Caspian Sea, and dispense with the present necessity of transhipping cargoes, in the open sea, to vessels of small draught in order to convey them to Astrakhan.

PARALLEL JETTIES AT OUTLET.

The second method adopted for deepening one of the outlet channels of the delta of a tideless river, consists in prolonging the channel beyond the coast-line by a jetty on each side, so that the current, instead of spreading out at the mouth, may be continued out to the bar, and increase the depth over it by scour. The advantage of this system is that the additional depth is maintained by the same scouring force which created it, instead of necessitating constant works of maintenance, such as are required in the case of improvements by dredging in channels exposed to sedimentary deposit. It is essential, however, for the success of this system, that the sea-bottom should shelve down beyond the bar; for the deposit removed by scour from the bar is only carried somewhat further out to sea by the prolonged current, and deposited when the current slackens beyond the ends of the jetties, except in so far as it may be partially borne away by any littoral current. If, therefore, the sea continues shallow beyond the bar, the effect of the jetties is merely to push the bar a little further out in proportion to their length; and this, in the case of the Volga jetty works, together with the leakage through the interstices of the jetties, and the increase of the volume of alluvium by the barring of the other outlets of the Kamysiak branch, explains the failure of these works.

The amount of permanence of the improvement in depth effected by parallel jetties is dependent upon four conditions;

namely, the slope of the sea-bottom beyond the bar; the density of the alluvium brought down; the power and extent of any littoral current or wave-action; and the rate of advance of the delta at the outlet selected. A good depth seawards defers the period when the accumulation of deposit further out will rise high enough to form a new bar. Alluvium of too great density to be carried in suspension is less affected by the concentrated current than the lighter materials, and therefore comes to rest again nearer the coast-line, and is brought less under the influence of any littoral current. A strong littoral current and powerful wave-action carry away some of the alluvium, and therefore delay the formation of a new bar beyond the jetty channel. The rate of advance of the shore in front of the outlets of any special delta is proportionate to the discharge, for similar depths of the sea in front; and this consideration affects the choice of the outlet, and is the only one of the four conditions which is subject to modification.

Rhone Jetties. In 1852, previously to the commencement of jetty works at the mouth of the Rhone, the Piémanson and Roustan branches (Plate 5, Fig. 1) had depths at their outlets of $9\frac{1}{2}$ feet a third of a mile inside the bar, and $5\frac{1}{8}$ feet over the bar; and the former had a depth of $33\frac{1}{2}$ feet a third of a mile beyond the bar, and the latter 63 feet at the same distance out. The depths at the Eugène outlet, at similar positions, were $10\frac{1}{2}$, $4\frac{1}{8}$, and $58\frac{3}{4}$ feet; and for the East outlet, $5\frac{1}{4}$, $3\frac{1}{8}$, and $13\frac{1}{2}$ feet[1]. Evidently, therefore, the Roustan outlet was the best for improvement, as it possessed as good depths up to, and over the bar, as the Piémanson outlet, and a much better depth seawards, a most important condition for a river charged with dense detritus: whilst the East outlet was in every respect much the worst as regarded depth. Nevertheless, this eastern branch was selected, on account of its more favourable direction for the entry and exit of vessels, a matter of little value if the depth is inadequate.

---

[1] 'Cours de Navigation intérieure.' M. Malézieux, p. 179.

Embankments were carried out along each side of the East channel in 1852–57, from above the entrances of the other branches to within half a mile of the bar (Plate 5, Fig. 1); and the three southern channels were thus shut off, as well as the two minor northern ones; and the whole discharge of the Rhone was concentrated in the East branch. The increased scour thereby produced at the eastern outlet removed the existing bar, and for a time materially deepened the outlet channel, increasing the average available depth from $4\frac{1}{2}$ feet in 1852, up to $9\frac{3}{4}$ feet in 1856. The bar, however, gradually formed again further out; and the average depth over it has never since attained the moderate increase in depth of 1856, having ranged between $8\frac{2}{3}$ feet in 1873 and $3\frac{3}{5}$ feet in 1890. The maximum depth since 1856, which results from great floods, reached $11\frac{1}{5}$ feet in 1859 and 1873; and the minimum depth of slightly under 2 feet occurred in 1890, during which year the depth never exceeded $6\frac{1}{2}$ feet. Three states of the bar, in 1841, 1873, and 1891 respectively[1], are shown in Plate 5, Fig. 9, from which it appears that the bar is a mile further out than fifty years ago, the main advance being due to the embankments, though a distinct advance since 1873 is also manifest. The three sections also show how small has been the improvement effected by the parallel jetty system as applied to the Rhone, even in an exceptionally favourable year such as 1873. The average depth, indeed, over the bar during the last thirty years amounts to only $6\frac{1}{8}$ feet, a gain of 3 feet over the depth of the eastern bar in 1852, but slightly less than the depth over the Eugène bar in 1841. Moreover, the bar having been pushed further out from the shore by the increased discharge, is in a more exposed situation. The steep sea-slope of the bar is the result of the large proportion of heavy sediment rolled along the bed forming the bar, which comes to rest before the

[1] Charts of the Rhone in 1841 and 1891 were furnished me by Monsieur A. Guérard, the engineer in charge of the maritime part of the river.

current passes the crest of the bar, and is only gradually pushed over on to the outer slope.

The failure of parallel jetties to effect an adequate deepening at the Rhone outlet is due to the course adopted, rather than to any inherent defect in the system, or specially unfavourable physical conditions. The waters of the Rhone, indeed, bring down a rather large proportion of sediment, and a notably large percentage of this sediment is heavy; but this unfavourable circumstance is somewhat compensated for by the existence of an oceanic littoral current extending to a considerable depth, and by the declivity of the bottom of the Mediterranean Sea along the southern front of the Rhone delta. The selection, however, of the eastern mouth as the navigable channel, and the closing of the other branches, were fatal to the success of the works. The eastern outlet faces the prevalent and strongest winds blowing from the sea, and it discharges into the shallow sheltered Gulf of Foz, away from the influence of the littoral current; and the limitation of the discharge to this single outlet, whilst increasing the scour, has concentrated the advance of the delta on to this part, the comparative shoalness of which increases the rate of progression. The advance, indeed, of the delta across the gulf already impedes the access of vessels to the ports of Bouc and St. Louis during northerly winds; and the reopening of the southern mouths to check this advance has been proposed. The southern shore of the delta, on the contrary, has been eroded in places under the action of the sea, since the cessation of deposit at this part owing to the closing of the southern outlets.

If one of the southern mouths of the Rhone had been selected for improvement by parallel jetties, preferably the Roustan branch, there is every reason to believe that the navigable channel might have been adequately deepened, in spite of the density of the sediment, owing to the good depth in front, and the eroding influence of the waves and littoral current. Moreover, by leaving the other entrances open, so as to spread the alluvium discharged over a wide area, the rate of advance of the delta would not have been accelerated, and the entrance to the Gulf of Foz would not have been imperilled [1].

Jetties at Sulina Mouth of Danube. The Kilia branch of the Danube possesses the best navigable channel through the delta; but the number of its mouths, and the small depth over their bars, the rapid advance of its delta owing to its large discharge, and the flatter slope of the sea-bottom there than further south, rendered it unsuitable for improvement (Plate 5, Fig. 2). The St. George's branch was preferred by Sir Charles Hartley, the engineer to the Danube Commission, owing to its having a much better navigable channel through the delta than the Sulina branch, on account of the greater depth of the sea in front of its outlet, and as its mouth is 20 miles nearer the Bosphorus, and more favourably situated for navigation than the Sulina mouth [2]. The Sulina mouth was, however, selected for provisional improvement in 1858, on account of its being at that time the only navigable outlet, and owing to its bar being only half the distance from the shore of the St. George's bar, and the cost of the works therefore much less. The jetties were carried out to a depth of 18 feet of water in 1858–61; and they were consolidated in 1866–71, owing to their success and the insufficiency of funds for improving the St. George's mouth. They converge from the shore to the edge of the outlet channel, and then extend in a parallel direction, 600 feet apart, out from the coast [3], beyond the site of the bar in 1857, the crest of which was

---

[1] 'Amélioration de la partie maritime des Fleuves, y compris leurs Embouchures.' L. F. Vernon-Harcourt, Vme Congrès International de Navigation intérieure, Paris, 1892.

[2] 'Description of the Delta of the Danube.' C. A. Hartley. Minutes of Proceedings Institution C. E., vol. xxi, p. 283.

[3] 'Description of the Delta of the Danube.' Sir Charles A. Hartley. Minutes of Proceedings Institution C. E., 1873, vol. xxxvi, p. 201.

only ⅝ of a mile beyond the Sulina mouth (Plate 5, Figs. 3 and 4). The jetties were formed in the first instance of a mound of rubble stone and pilework, and consolidated subsequently by a concrete wall along the top, and random concrete blocks, from 10 to 20 tons in weight, on the sea-slope of the mound. The south jetty, which originally was overlapped by the north jetty, was extended 457 feet in 1869, and 204 feet more in 1876–77, so as to terminate at the same distance out as the other jetty, and thus scour away the sand which was heaped up against the inside of the projection of the north jetty under the influence of south-westerly winds. The shore end also of the north jetty was gradually extended landwards, between 1867 and 1883, a total length of 1,100 feet, to maintain its connexion with the shore which has been eroded by the waves and littoral current to the north of the jetties, owing to their projecting, like a groyne, about a mile out from the coast. These jetties directed the scour of the river across the bar, and pushed its crest out seawards beyond the pierheads, so that the navigable depth over it was gradually increased from 9 feet in 1857, up to 20½ feet in 1872, which depth has been since maintained (Plate 5, Fig. 10). The total cost of the works amounted to about £197,000, and the maintenance during the last twenty years has averaged about £2,300 a year.

The shelter afforded by the jetties on their southern side from the waves raised by northerly gales, and from the littoral current flowing from the north, has led to the deposit of alluvium brought in by south-westerly winds, producing an advance of the shore and of the lines of soundings to the south of the jetty channel. These accumulations deflect the deepest channel towards the north outside the pierheads, so that sometimes the depth falls below 18 feet in the direct line seawards of the jetty channel [1].

---

[1] 'The Sulina Mouth of the Danube.' C. H. L. Kühl. Minutes of Proceedings Institution C. E., 1888, vol. xci, p. 331, and plate 3, fig. 2.

The works were expected to procure a depth of 16 feet at the outlet; and therefore the results achieved were 4 feet in excess of the depth anticipated. Moreover, the navigable depth of $20\frac{1}{2}$ feet has been maintained for twenty years without any extension of the jetties, with the exception of the small addition to the southern jetty in 1876–77 in completion of the original design. The success attained is due to the lightness of the alluvial matter brought down, and its removal on being brought under the influence of the littoral currents by the action of the jetties in conveying the discharge seawards. The lightness of the alluvium is indicated by the comparatively small proportion of material deposited within $1\frac{1}{8}$ miles of the pierheads, amounting to less than one-sixth of the material carried in suspension. Moreover, a still larger proportion of the suspended matter must be conveyed by the currents beyond this limit before being deposited, as some of the deposit measured must consist of heavier sediment rolled along the bottom, the volume of which has not been ascertained. The form of the bar also, and the great contrast it presents on its sea-slope to the Rhone bar (compare Figs. 9 and 10 in Plate 5), show that a considerable portion of the sediment passes over the crest of the bar before being deposited by a kind of sifting process according to its density, and that, therefore, a large proportion of the sediment forming the bar is fairly light. The moderate ratio, moreover, of the sediment to the discharge as compared with that of the Rhone, and the smaller density of the Black Sea than of the Mediterranean, owing to the smaller proportion of salt in the former, are conditions favourable to the maintenance of the depth at the Sulina mouth. The comparatively small discharge also through the Sulina branch, and the non-interference with the discharges through the other outlets, retard the rate of shoaling in front of the Sulina jetties, and postpone the necessity for their extension. In fact the slope of the sea-bottom is the only condition which is less favourable at the Sulina mouth than at the mouth of the Rhone; whilst the errors committed in the selection of the outlet, and in closing the other outlets at the Rhone delta, have been avoided at the Danube delta.

Notwithstanding the favourable conditions under which the jetty system has been carried out so successfully at Sulina, these works must not be regarded as securing an absolutely permanent maintenance of the existing navigable depth. Though the depth of 20 feet attained in 1872 has hitherto been maintained, the sections over the bar in 1873 and 1891 (Plate 5, Fig. 10) show that deposit is extending on the outer slope of the bar; and the 4- and 5-fathom lines of soundings have progressed seawards a little over a quarter of a mile, in the direct prolongation of the jetty channel, between 1871 and 1891 (Plate 5, Fig. 3), amounting to an average advance of 70 feet a year; whilst the gradual advance of the shore-line to the south of the jetties is clearly marked [1]. The state of the sea-bottom in front of the Sulina jetties may present fluctuations resulting from variations in the volume of the floods of the Danube, the amount of sediment brought down, and the direction of the winds, occasionally reverting to a more favourable condition, as in 1891 compared with 1886; but eventually, sooner or later, the accumulation of deposit must gradually creep in front of the line of the jetty channel, and the depth in the navigable channel be at last reduced below 20 feet. A moderate prolongation, however, of the jetties will suffice to scour the bar again, and driving it into deeper water, secure the requisite navigable depth in the outlet channel for another period of years.

The outlet channel having had its depth more than doubled over the bar, afforded a much better depth than many parts of the Sulina branch at a low stage of the river. Accordingly,

---

[1] Sir Charles Hartley supplied me with a Chart of the Danube of 1891, from which the lines of soundings for 1891 on the plan, and the section of the bar in 1891, have been obtained.

the available depth throughout this branch has been improved by groynes to reduce excessive widths, training banks, easing bends, dredging shoals, and making cuts across very bad bends ; one of these cuts, a little over 5 miles in length in place of nearly 10 miles of winding channel, having been executed in 1890–94 [1] (Plate 5, Fig. 2). These works have already increased the minimum depth in the Sulina branch, at the lowest stage of the river, from 7 feet in 1857 to 13 feet in 1871, and 16¾ feet in 1891.

The tonnage of vessels outward bound with cargoes through the Sulina mouth, which averaged about 300,000 tons in 1857, rose directly after the improvement of the channel, and reached nearly 1½ million tons in 1889.

Mississippi South Pass Jetties. The South-West Pass was proposed by Mr. Eads for improvement by parallel jetties at its outlet, on account of its possessing a larger channel through the delta than the South Pass, and owing to the South Pass being impeded by a shoal at its head. Eventually, however, the smaller South Pass was chosen for improvement, as in the case of the Danube, owing to the much shorter jetties needed for directing the currents out to the bar at this pass than at the South-West Pass, whose bar was twice as far off from the shore [2].

The works at the outlet of the South Pass consist of two parallel jetties about 1,000 feet apart, formed of long willow mattresses consolidated with limestone, and with large concrete blocks at their outer ends (Plate 5, Figs. 6 and 8), carried out in 1876–79, for lengths of 2¼ and 1½ miles on the east and west sides of the channel respectively, from the shore out to the bar. They terminate, at the same distance out, in 30 feet of water, and were slightly curved round towards the south at their extremities, to direct the discharge of the river at their

---

[1] 'The Sulina Branch of the Danube.' C. H. L. Kühl; Minutes of Proceedings Institution C. E., 1891, vol. cvi, p. 239.

[2] 'A History of the Jetties at the Mouth of the Mississippi River.' E. L. Corthell.

outlet at right angles to the littoral current flowing from east to west across the ends of the jetties. The jetty channel has been further regulated, and the scour along the central portion increased, by the addition of an inner line of training banks of willow mattresses on each side (Plate 5, Fig. 6). The concentrated current at the outlet produced by these works soon scoured out the channel between the jetties, and lowered the bar beyond; so that by 1880, an outlet channel had been formed with a minimum central depth of 31 feet throughout; whereas in 1875 there was a depth of only 8 feet over the bar (Plate 5, Fig. 7).

The shoal at the head of the South Pass, where it leaves the main river, which had a depth of water of only 15 feet over it, was lowered, in 1876–77, by contracting the funnel-shaped entrance to the South Pass, by mattress dykes on each side, into a uniform channel 850 feet wide. These dykes concentrated the current entering the pass, over the shoal; and after dredging a narrow channel through the shoal to a depth of 18 feet to hasten the deepening, the floods of 1877 scoured out a good channel; and by 1880 a depth of 30 feet had been attained across the shoal. As these dykes, by narrowing the entrance to the South Pass, and therefore somewhat checking the influx from the river into this pass, would have diverted a portion of the discharge into the other passes, mattress sills, 70 feet in width, were sunk across the beds of the entrances to the South-West Pass and the Pass à l'Outre, so that by slightly reducing the sectional areas of these channels, the influx into them might be similarly checked, and the proportionate discharges into the three passes might remain unaltered.

The concentration and prolongation of the current at the outlet by means of the parallel jetties, not merely lower the bar, carrying the material scoured away, and the sediment constantly brought down, into deeper water, but also bring the discharged alluvium more under the influence of the stronger littoral current further out from the shore, till the gradual advance of the shore-line at the back of the jetties shall have pushed the littoral current further out into the gulf. Though, however, the lighter alluvium may be carried away by the littoral current, the heavier material is merely transported into the deeper water beyond the ends of the jetties, where shoaling is gradually taking place (Plate 5, Fig. 7). In fact, within a fan-shaped area of $1\frac{1}{4}$ square miles in front of the outlet, the decrease in depth between 1876 and 1892 has amounted, on the average, to over 13 feet over the whole area [1], the shoaling having reached a maximum of $2\frac{7}{10}$ feet in the year 1890–91. Altogether, the greatest amount of shoaling has occurred in the outer zone of this area, from 4,000 to 6,000 feet beyond the ends of the jetties; but in 1890–91 and 1891–92 the shoaling was greatest in the central zone. The least advance of the lines of soundings beyond the ends of the jetties, between 1877 and 1892, occurred at the 20-foot line, amounting to an average of 810 feet, or about 54 feet annually, the advance of the 30-foot line averaging 63 feet annually, but exhibiting the maximum advance of 240 feet in 1891–92. Further out the rate of advance increases, and reaches a maximum at the 70-foot line, where the average advance was 1,555 feet from 1877 to 1892, or 104 feet a year. The rate of advance decreases in greater depths, beyond the limits of the area of $1\frac{1}{4}$ square miles; but even up to the limit of the soundings, at the 100-foot line, the annual progression is large, attaining an average of 87 feet at the 80-foot line, 83 feet at the 90-foot line, and 81 feet at the 100-foot line. A new bar, accordingly, is in process of formation about a quarter of a mile beyond the ends of the jetties; so that in 1891, the depth in the prolongation of the central line of the jetty channel had been reduced to a minimum of 15 feet. The deep outlet channel also, which still maintains

[1] 'Annual Report of the Chief of Engineers, U.S.A., for the year 1892,' part 2, p. 1478.

a minimum depth of 30 feet, had been deflected to the east by the accumulation of deposit on the western side of the outlet; and on emerging from between the jetties, diverged about 500 feet to the east of the east jetty, and was very tortuous and narrow in places. Dredging has consequently been resorted to for forming a better and more direct channel (Plate 5, Fig. 6). During eleven days, moreover, in 1891, and for a month in 1892, the central depth in the jetty channel fell below the statutory depth of 30 feet. It is evident, therefore, that though the South Pass jetties have hitherto maintained a channel of 30 feet in depth at the outlet, this channel can hardly even now be described as suitable for navigation; dredging in front of the jetties will become increasingly necessary; and there is every prospect that the accumulation of deposit taking place in front of the outlet, will before long necessitate the prolongation of the jetties, to make the deep channel more direct, and to maintain a navigable channel of 30 feet at the outlet for another period of years.

Mississippi and Danube Jetty Works compared. The jetty works at the mouth of the Mississippi have been carried out on precisely the same principles as the Danube jetty works; though the greater size of the South Pass, the longer distance of its bar from the shore, and the deeper channel required for the trade of the Mississippi, have necessitated more extensive jetties than at Sulina (Plate 5, Figs. 3, 6, 7, and 10). The outlet of the smallest branch through the delta was selected for improvement in both cases; parallel jetties were carried out to the bar; and the discharge through the other branches was not interfered with. The results also of the works have been very similar: the bar in both cases has been pushed out seawards by the concentrated current, and the depth over it was at once increased, and has since been fairly maintained; whilst the influence of the littoral current on the alluvium has, in both instances, caused a diversion of the deepest channel away to one side from the direct line of the jetty channel; and the advance of the lines of soundings seawards, owing to the gradual deposit of the alluvium of the river in front of the outlet, has been most marked at some distance beyond the ends of the jetties. The advantage of the sea-slope in front of the South Pass being four times steeper than in front of the Sulina mouth, has been neutralised by the sediment brought down by the Mississippi being much greater in proportion to the discharge, and also much denser than the alluvium of the Danube. Moreover, the absence of the erosion by the sea along the shore of the South Pass, which has produced an increase in depth to the north of the Sulina jetties, combined with the unfavourable character of the alluvium, and the necessity of maintaining a minimum navigable depth of 30 feet at the mouth of the Mississippi, render the prospect of a prolongation of the jetties more imminent at the South Pass than at Sulina.

SHIP-CANAL INSTEAD OF A DELTA CHANNEL.

Owing to the natural impediments to navigation existing at the outlets of a delta, it has sometimes been proposed to connect the river, above its delta, with the sea by means of a ship-canal. This plan resembles in principle the lateral canals which are resorted to for forming a connecting link between the two navigable portions of a river, in consequence of the intervention of rocky rapids, or the existence of an insurmountable obstacle such as the falls of Niagara. The new outlet must be situated beyond the influence of the alluvium of the river, otherwise it would become silted up. Thus the deep-water entrance to the Tiber, provided in the reign of the Emperor Claudius, by the construction of the port of Ostia, and connecting it with the river by a canal, was within the range of the alluvial deposits at the mouth of the

Tiber; and consequently the port gradually silted up, and is now two miles from the coast.

In a few instances, a ship-canal has been formed, diverging sufficiently from the delta to provide a deeper permanent outlet for the navigation of a tideless river than the channels through the delta can afford. Thus the San Carlos Canal, starting from the river Ebro at Amposta and emerging into the Bay of Alfaques to the south of the delta of the Ebro, provides a navigable channel for the trade of the river, which is cut off from its natural outlet by the impediments of the delta projecting into the sea, and barred by accumulations of sand. The Mahmoudieh Canal leaves the Rosetta branch of the Nile at Atfeh, about 30 miles from the Rosetta mouth; and after a westerly course of 50 miles, it enters the harbour of Alexandria. This canal, which cost £300,000, and was opened in 1820, is only 100 feet broad and of a very moderate depth; but it affords a deeper and much more sheltered outlet for the traffic of the Nile than either of the shallow and exposed mouths provides. In this case, the great variation in the discharge of the river at high and low Nile, and the paramount importance of irrigation works for Egypt, preclude any attempt to improve one of the delta channels.

The construction of a ship-canal for avoiding the obstacles of the Volga delta has been considered, but has been abandoned, as the canal would have to be given a length of 125 miles in order to provide Astrakhan with an outlet to the Caspian Sea beyond the influence of the delta; and the cost of such a work would be prohibitive.

A ship-canal was one of the proposals submitted for improving the navigable outlet of the Danube, diverging to the south of the St. George's branch, so as to avoid the shoals in front of its mouths; but the improvement of the Sulina branch by parallel jetties was fortunately preferred, as it has provided an unimpeded and deeper channel formed by natural scour.

The formation of a ship-canal for providing a deep-water outlet for the Mississippi was advocated by several persons in opposition to the jetty system; but eventually the advocates of the jetty system prevailed, with the result that an unimpeded outlet with a greater depth than proposed for the canal has been obtained.

Conclusions. Harrowing on the bars of the Danube and the Mississippi failed to produce any material or permanent improvement; and owing to the inability of the issuing current to transport its own burden of sediment, the increase of its charge by stirring up the deposit offers no prospect of securing any adequate increase in depth.

Dredging proved inefficacious to secure an adequate deepening of the outlet channels of the Danube and the Mississippi; and it is evident that this method of improvement would have to be carried out continuously on a very large scale to cope with the great volume of alluvium brought down by such rivers. It is proposed, indeed, to adopt this system of improvement in the case of the Volga; but the small depth in front, and the absence of any littoral current preclude the use of the jetty system in this instance; and the increase in depth anticipated is small. Dredging alone, accordingly, appears chiefly applicable to the improvement of the outlets of tideless rivers when, owing to unfavourable local conditions, the jetty system cannot be resorted to, or in cases where the volume of alluvium brought down is comparatively small. Dredging is to be commenced very soon at Sulina, with a bucket-ladder hopper dredger of large dimensions provided with a sand pump, in order to endeavour to increase the depth of the entrance channel of the Danube across the Sulina bar to 23 or 24 feet. These operations will be of considerable interest, in showing to what extent dredging can be employed for increasing the improvement effected at an outlet by jetties, and also, eventually, in proving whether dredging is economically capable of dispensing with, or at any rate of deferring for several years, the prolongation of the jetties for maintaining the depth. Dredging is likely to prove more applicable at the Sulina mouth than at the South Pass outlet of the Mississippi, owing to the smaller volume of alluvium brought down by the Sulina branch, and the smaller depth of the sea in front.

The jetty system has produced most successful results at the Sulina mouth of the Danube and the South Pass of the Mississippi; and the absence of improvement at the Rhone outlet is fully accounted for by the unfavourable direction of the outlet selected, the concentration of the discharge into a single mouth by closing the other mouths, the absence of a littoral current in front of the trained outlet, and the density of the sand brought down. Parallel jetties may, indeed, be regarded as furnishing a satisfactory method of improving the outlet channel of tideless rivers, provided the mouth selected is suitably situated, with deep water in front and a littoral current sweeping across it, and the other mouths are not closed. The volume of material discharged, and its density, affect the results obtained; and the cost is reduced if one of the smaller channels, just adequate for navigation, is selected, where the progress of the delta is slower and the distance out of the bar is less. Though the system does not constitute an absolutely permanent improvement, the jetties generally suffice to maintain the depth for several years, till a bar has had time to form further out; and eventually an extension of the jetties would ensure the maintenance of the depth for another period of years.

The evasion of the difficulty at the outlet, by the construction of a ship-canal with its outlet beyond the influence of the delta, should not be resorted to unless the jetty system is inapplicable, as it generally would constitute a work of considerable magnitude to secure an adequate depth, its passage is necessarily impeded by a lock to shut out the turbid river current, and any increase in depth at its outlet has to be procured by dredging instead of by the natural scour.

CHAPTER IX.

JETTIES AND BREAKWATERS AT THE MOUTHS OF RIVERS.

Bars: fluvial, marine; origin of; favourable and unfavourable conditions. *Jetties at Outlets of Tideless Rivers free from Silt:* Swinemünde, Vindau, Dwina, Pernow, Chicago, Buffalo, and Oswego. *Converging Breakwaters at Mouths of Tidal Rivers:* objects; Liffey, Merrimac, Tees, Tyne, Wear, Nervion; Remarks. *Training Jetties at Mouths of Tidal Rivers:* objects and instances; Yare, Adur, Adour; combined with training works on Maas up to Rotterdam, and on Nervion up to Bilbao. Remarks on works at Outlets: deepening aided by dredging; materials for jetties; alteration of outlet.

Bars. The rivers described in the preceding chapter themselves form the bars at the outlets of their delta channels, by the alluvium which they bring down from inland. The bars, however, found at the mouths of tideless rivers which are not in the process of forming deltas, owing to the small amount of material they carry into the sea, and at the mouths of tidal rivers in which the tidal volume is much larger than the fresh-water discharge, cannot be attributed to this cause. In the case of these latter rivers, the bar is of marine, and not of fluvial origin, and is the result of the heaping-up action of the waves along the coast, tending to form a continuous beach across the mouth of the river, which, in most instances, is only partially prevented by the scour of the tidal ebb and flow, and by the fresh-water discharge of the river. The river Neva is an instance of a tideless river which has a marine bar at its mouth; for though it brings down a large quantity of alluvium in its course, it deposits it all in Lake Ladoga, thirty-six miles above its outlet into the Gulf of Finland, an arm of the Baltic Sea. The same is the case with the outlets of the Oder into the Baltic, for this river, like the Neva, though deltaic in appearance, does not at the present time discharge much alluvium into the sea, as its burden of sediment is deposited in some lakes and the Stettiner Haff before reaching the sea. The outlet channels, also, of the Rhine and the Maas, which, discharging a large volume of water charged with alluvium into a part of the North Sea where the rise of tide is small, have formed a regular delta in ancient times, must now be classed as tidal rivers, since the sea tends to encroach upon the land along that coast, and prevents any advance of the foreshore.

At the mouths of rivers with a large tidal flow, the marine origin of the bar is evident, both from the nature of the materials of which it is composed, and also on account of the inadequacy of a moderate fresh-water discharge to supply the large volumes of detritus encumbering many large tidal estuaries.

The term bar is generally applied to the ridge or kind of submerged beach which intervenes between the mouth of the river and deep water, stretching out seawards in front of the outlet in the line of the greatest scour of the river, and curving round to form a junction with the foreshore on each side, as very clearly defined in the chart of the Mersey estuary (Plate 7, Fig. 11). The outer, lowest part of the ridge, in the line of the deepest channel, is actually called the bar, for it is this portion, which being higher than the navigable channel on either side, impedes the passage of vessels of large draught between the river and the sea. A distinct hollow is found in most rivers just inside the bar, which must be due to the action of waves and of the flood tide in crossing the bar (Plates 6, 7, 8, and 9).

Shoals often exist across the navigable channel in an unimproved river, over which the depth of water is as little, or even less than over the bar outside; but being fairly sheltered, they can generally be lowered without great difficulty by dredging or training works. The bar, on the contrary, being fully exposed, and having the available depth over it reduced by the wave oscillations during storms, presents a more serious danger to navigation, and is far more difficult to improve than inner shoals. The bars, however, at the mouths of tideless rivers free from alluvium, and of tidal rivers, when once adequately trained or protected by jetties or breakwaters, are not liable to form again beyond the extremities of the works after a certain lapse of time, as in the case of delta-forming rivers ; but the principles on which the works at the mouths of tidal rivers should be designed are more complex than in tideless silt-bearing rivers, on account of the variability in the physical conditions, due to differences in tidal influence.

The height and size of the bar are proportionate to the amount of drift brought along the coast, and to the force and duration of the prevailing winds and littoral currents which occasion the accumulation of a portion of this drift across the mouth of a river. An exposed site, and deep water near the shore favour the formation of the submarine beach, by increasing the force of the waves ; but a sheltered position and outlying sandbanks, whilst promoting deposit, at the same time reduce the transporting power of the waves. A large tidal ebb and flow in a river, and a large fresh-water discharge serve, on the contrary, to lower the bar, and to drive the drift into deep water. When the influence of these latter favourable conditions is small, and the force of waves beating obliquely on the coast and of littoral currents is considerable, the natural mouth of a river is sometimes closed up by drift ; and the river is obliged to find an outlet at another part of the shore. Instances of this occurrence are afforded by the river Yare on the Norfolk coast, the river Adur on the Sussex coast, and the river Adour on the south-west coast of France near Bayonne.

JETTIES AT OUTLETS OF TIDELESS RIVERS FREE
FROM SILT.

In the absence of any tidal rise, the only means available
for lowering the bar at the mouth of a river consist in the
scour of the current of the river, directed by jetties across the
bar, or in dredging a channel through the bar.   The improvement by jetties, under favourable conditions, is fairly permanent, for it extends the scouring power of the discharge
of the river across the foreshore into deep water, and thus
forms a channel through the submarine beach, and prevents
its filling up again with drift under the action of the sea.   The
system, indeed, is similar to that adopted at delta outlets,
with the advantage that, in the absence of alluvium discharged
by the river, the bar does not gradually form again further
out.   The deepening of the outlet channel by dredging is not
so permanent as that produced by the regulated scour of the
river ; but it is simple, and can be gradually extended as the
needs of any ports on the river, or the increase of trade may
render expedient.

Swinemünde Jetties.   The Swine mouth of the Oder, traversing the neck of land separating Stettiner Haff from the
Baltic Sea, forms the navigable outlet of the Oder, and the
entrance to the port of Swinemünde.   This mouth was formerly
obstructed by a bar, over which the available depth did not
exceed 7 feet.   In 1818–28 jetties were constructed curving
round to a northerly direction, and prolonging the channel
for about a mile across the foreshore, in the line followed
by the issuing current of the river (Fig. 13, p. 206).   These
works, by concentrating the scour across the bar, deepened
the channel over it to 28 feet ; but this depth was eventually
reduced, owing to the drift brought into the channel from the
west, round the shorter west pier, by the prevailing north-
westerly winds and the littoral current flowing from west to
east.   Dredging has been resorted to for maintaining the

depth, which, moreover, is favoured by the concave form of
the east jetty towards the channel, so that in spite of the
exposure of the outer part of the jetty channel to the drift
from the west owing to the shortness of the west jetty, the
channel at the present time has an available depth of 22½ feet.

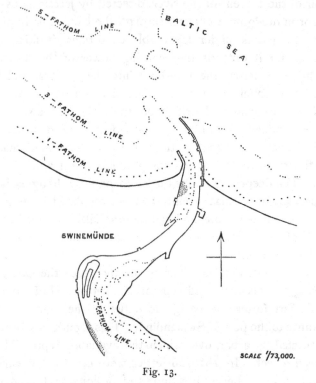

Fig. 13.

Jetties at Swine Mouth of River Oder.

In addition to the dredging, the east jetty has been prolonged
about 500 feet, to extend its guiding influence on the navigable channel which keeps close alongside it ; whereas the
channel between the jetties shoals from 18 feet in mid-channel,
to under 6 feet against the west jetty. In this case, it is only the
peculiar concave form of the east jetty, following the natural
course of the outlet, which has enabled a prolongation of the

western jetty, to correspond with the east jetty, to be dispensed with.

Russian River Outlets into the Baltic. The entrances to some Russian ports near the mouths of rivers flowing into the Baltic Sea, have also been improved by jetties, in prolongation

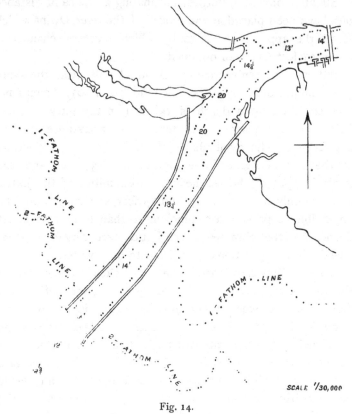

Fig. 14.

Jetties at Mouth of River Pernow.

of the banks of the river, concentrating the current across the bar. Thus the mouth of the river Vindau, giving access to the port of Vindau, has been deepened to 18 feet, by carrying out jetties on each side into the sea, the southern jetty in prolongation of the concave bank of the river having been extended beyond the northern jetty, as in the case of the Swine; though, in this instance, the southern jetty not merely guides the issuing current which follows the concave shore, but also protects the channel from the prevalent south-westerly winds and the main drift along the coast [1].

Slightly converging jetties, terminating at the same distance out, have been placed at the mouth of the river Dwina which forms the approach to Riga, by which a central channel of about 21 feet has been obtained.

Jetties have been carried out about $1\frac{1}{4}$ miles from the shore at the mouth of the river Pernow (Fig. 14, p. 207), which have secured a minimum depth of $13\frac{1}{2}$ feet in the jetty channel, falling in one place to 12 feet a short distance beyond the jetties. The available depth, however, in the river up to the port of Pernow does not exceed $13\frac{3}{4}$ feet, and the general depth of the sea near the extremities of the jetties amounts only to between 13 and 15 feet, so that it would be impossible to provide a channel more than from 1 to 2 feet deeper than that obtained by the jetties, except by very extensive dredging operations carried out to deep water.

River Outlets into North American Lakes. The outlets of the river Chicago into Lake Michigan, the river Buffalo into Lake Erie, the river Oswego into Lake Ontario, and of other rivers flowing into these inland lakes, have been trained and deepened by parallel cribwork jetties filled with rubble stone, the increase in depth being aided by dredging. The windward jetty has generally been made to overlap the other, in order to keep the drift along the coast from being driven into the outlet channel; and sometimes the foreshore has advanced beyond the jetties, necessitating their extension. Some years ago, however, cribwork breakwaters were commenced at the mouths of the above-mentioned rivers, enclosing a considerable space to afford increased accommodation for vessels, which also serve

[1] 'Atlas des Ports de Commerce de la Russie,' part 1.

to shelter the outlet of the river, and place the entrance to the port in deep water [1].

<h2 style="text-align:center">CONVERGING BREAKWATERS AT THE MOUTHS OF TIDAL RIVERS.</h2>

The convergence of jetties, to increase the scour, is not admissible in the case of tidal rivers; for by narrowing the outlet, the influx of the flood tide would be checked, and the scouring power of the ebb correspondingly reduced. Converging breakwaters, however, starting from the shore at some distance · apart, and leaving a sufficiently wide opening at their extremities to allow of the free admission of the flood tide into a river, have been adopted with advantage; for the area enclosed by these breakwaters forms a kind of sluicing basin which, being filled by each flood tide, produces a tidal scour on the ebb through the narrowed outlet between the pierheads; and the depth is thereby increased, as exemplified by the works at the mouth of the river Liffey (Plate 6, Figs. 1 and 2). Generally, the converging breakwaters serve also to shelter the outlet of the river, and to protect it from the inroad of drift brought along the coast by waves and littoral currents, as, for instance, the breakwaters at the mouth of the river Tees (Plate 8, Fig. 4). Sometimes, moreover, the breakwaters enclose a large enough area of sufficient depth to provide a sheltered approach to the river, and a harbour of refuge, in which dredgers can easily remove the bar and deepen the navigable channel, of which the works at the mouth of the river Nervion, and of the river Tyne, afford examples (Plate 6, Fig. 12, and Plate 8, Fig. 1).

Breakwaters at Mouth of River Liffey. The river Liffey, constituting the entrance to the port of Dublin, flows into Dublin Bay across a sandy foreshore, about $2\frac{1}{2}$ miles wide, formed by the gradual accumulation of sand brought in from the sea, which dries at low water. Formerly the outlet channel

---

[1] 'Harbours and Docks.' L. F. Vernon-Harcourt, plate 5, figs. 5, 11, and 12.

was constantly shifting, and was obstructed by a bar in front of its mouth[1]. In 1711 the improvement of the river below Dublin was commenced, by rectifying its channel, and lining it with quays ; and during the progress of these works, the Great South Wall was begun (Plate 6, Fig. 1), with the object of straightening the outlet channel, and protecting it from south-westerly gales and the inroad of sand from the southern strand. This breakwater, 3⅛ miles long, was only completed in 1796 ; but though it fulfilled the purposes for which it was designed, and also somewhat deepened the channel at the mouth of the river by the tidal scour produced round its head, it had no effect in lowering the bar beyond. Accordingly, a second breakwater, called the Great North Wall, formed of rubble stone, and converging from the shore towards the south breakwater, was carried out in 1820-25 (Plate 6, Figs. 1 and 3). This breakwater, 1¾ miles long, was only raised to half-tide level along its outer portion, which facilitates the influx of the flood tide into the enclosed area, and prevents the occurrence of rapid currents through the contracted opening between the heads of the breakwaters, 1,000 feet in width, near high water, which would be inconvenient for navigation. The tidal scour, moreover, is concentrated between the pier-heads during the latter half of the ebb, when most efficacious for lowering the bar, and when, owing to the peculiar set of the tidal currents in the bay near low water, there is no danger of the sand removed during the latter part of the ebb being brought back again into the sheltered area by the flood tide.

The river Liffey has a small discharge, and therefore the outlet channel below Dublin has had to be improved by dredging, the outer portion, across the foreshore, being protected by the breakwaters. The channel, however, across the bar outside has been deepened by the tidal scour produced by the emptying of the large area enclosed by the breakwaters, covered by water at each tide, the rise of tide in

[1] Minutes of Proceedings Institution C. E., vol. lviii, p. 119.

Dublin Bay at springs being 13 feet. The tidal capacity of this area has been gradually increased, owing to the washing of sand from the north strand by the flow during the ebb into the river channel, which is periodically removed from the channel by the dredgers, and also by dredging ballast from the north strand; whilst reclamation within this area has been strictly prohibited, in order that the tidal scour at the outlet may be fully maintained. This scour has gradually lowered the bar in front of the outlet, from a depth of 6 feet in 1819 to 16 feet in 1873, below low water of spring tides, which depth has been since maintained (Plate 6, Fig. 2); and this depth has been extended up to the port by dredging [1]. The improvement of 10 feet in depth effected by tidal scour will be maintained so long as the tidal capacity of the foreshore within the breakwater remains undiminished, and till the gradual advance of the foreshore outside the breakwaters brings low-water mark down to the ends of the breakwaters, when an extension of these works will become necessary.

Converging Jetties at Mouth of Merrimac River. The mouth of the river Merrimac is contracted from a wide expanse of flats inside, to a width of 1,000 feet with a depth of 30 feet. The scour through this opening, from the outflow of the tidal waters covering these flats, with a rise of tide of $7\frac{1}{2}$ feet, maintains the depth at the outlet; but on entering the Atlantic Ocean, the current soon loses its efficiency, and a bar due to wave-action existed across the outlet channel, with a depth of only 7 feet over its crest at mean low-water. Rubble jetties starting from the shore on each side, 4,000 feet apart, and converging to a width of 1,000 feet over the bar, were commenced in 1881. These jetties are designed to be 4,000 and 2,400 feet long respectively. By the middle of 1892, the north jetty had been carried out 2,485 feet, and the south jetty 1,077 feet, their extremities being only about

---

[1] Mr. J. P. Griffith furnished me with a recent chart of Dublin Bay, and other particulars.

1,300 feet apart; so that little more than the parallel outer portions, 1,000 feet long, remained to be constructed[1]. The concentrated scour across the bar, produced by these works, has already increased the depth over the bar to 13 feet at mean low water; and it is expected that, on the completion of the works, a channel across the bar, 17 feet deep, will be obtained by the tidal scour. In this case, such extensive works as those carried out at the mouth of the Liffey were not required, for a natural sluicing basin already existed inside the mouth, supplying the tidal reservoir for scour, which had to be created artificially in Dublin Bay.

Breakwaters at Mouth of River Tees. The South and North Gare breakwaters, converging together from the opposite shores at the outer extremity of the wide Tees estuary, shelter the estuary and the training walls, protect the enclosed area and the trained channel from sand brought along the coast, and when completed will aid in maintaining the outlet channel across the bar (Plate 8, Fig. 4). The South Gare breakwater, commenced in 1863, formed of a mound of the refuse slag from the neighbouring ironworks, protected on its sea face and round its head by a concrete wall, has been completed[2]; it stretches across the Bran Sand to the bar, and has a length of nearly 2½ miles (Plate 8, Fig. 6). The North Gare breakwater is in progress, also formed of a mound of slag, which is to be carried out for a length of rather over a mile, across the sands to the bar, leaving an opening between its extremity and the end of the other breakwater of 2,400 feet. The rise of spring tides at the mouth of the Tees is 15 feet, so that, on the completion of the North Gare breakwater, there will be a considerable flow between the pierheads at certain states of the tide, owing to

---

[1] 'Annual Report of the Chief of Engineers, U.S.A., for the year 1892,' part 1, p. 551; and atlas, plate 2.

[2] 'River Tees Improvements.' John Fowler; Minutes of Proceedings Institution C. E., 1887, vol. xc, p. 344.

the wide tidal area enclosed. This current will serve to maintain the depth at the outlet; but if sand is introduced by the flood tide during storms, and if reclamations are continued inside the estuary, the tidal capacity will be reduced, and the scour at the outlet correspondingly diminished. The increase in depth of about 10 feet since 1852, already obtained over the bar at the mouth of the Tees, has been effected by the scour resulting from the training works through the estuary, and dredging, aided of late years by the guidance and protection of the South Gare breakwater (Plate 8, Fig. 5). The converging breakwaters will always shelter the trained channel, will protect the channel and estuary from the inroad of drifting sand, and, by placing the outlet in comparatively deep water, will reduce considerably the amount of sand liable to be brought into the sheltered area by waves, where it would readily deposit; but the efficiency of the breakwaters in aiding the maintenance of the outlet channel by scour, will depend on the relinquishment of further reclamations in the estuary, and the absence of deposit in the enclosed tidal area.

Breakwaters at Mouth of River Tyne. The breakwaters starting from the sea-shore on each side of the mouth of the river Tyne, have been extended out beyond the bar into a depth of 5 fathoms at low water, and converge so as to leave an opening of 1,300 feet between the pierheads[1] (Plate 8, Figs. 1 and 3). The rise of spring tides at the mouth of the river is 14¾ feet; and the harbour entrance has been left adequately wide to admit the flood tide freely up the deepened river. The increased depth over the bar, of about 18 feet (Plate 8, Fig. 2), has not in this case been produced by any scouring effect of the converging breakwaters, but is due to dredging and the improved tidal scour resulting from the great deepening of the river above. The breakwaters, however, have enabled the bar to be dredged

[1] 'River Tyne Improvement.' P. J. Messent, 1888.

away under the shelter they have afforded; and they have also prevented the drift brought along the coast by the prevalent north-easterly winds, and by the flood tide, from encumbering again the mouth of the river, and have stopped the waves from re-forming the bar across the outlet. The north and south piers, as the breakwaters are called, were commenced in 1856, and, being extended as funds permitted, have only been quite recently completed, with total lengths of about 3,200 and 5,400 feet respectively. As the river flows directly into the North Sea without any intervening estuary, and the bar was only a short distance beyond the mouth of the river, smaller works would have sufficed for merely improving the outlet channel; but shortly after the commencement of the piers, it was decided to extend them sufficiently to form a harbour of refuge, in view of the unprotected state of the east coast of Great Britain between the Humber and the Firth of Forth.

Breakwaters at Mouth of River Wear. Till the south entrance to the Sunderland docks was built, the river Wear afforded the only approach to the port. During the eighteenth century, jetties were built on each side of the mouth of the river, to direct the out-flowing current across the bar which was raised by drift from the north during north-easterly storms, reaching nearly up to low-water spring tides[1] (Plate 6, Figs. 6 and 7). Though the drift naturally produced an advance of the foreshore against the northern jetty, the outlet channel was deepened to 4 feet at low water by the scour over the bar occasioned by the trained outlet, aided by dredging and the removal of shoals; and the jetties, which had originally been built of timber, were rebuilt with stone about 1843. The rise of tide at the mouth of the Wear is 14½ feet at springs; and the scour at the outlet is aided by tidal recesses which have been retained near the mouth, and by the discharge from a hilly drainage-area of 456 square

[1] Minutes of Proceedings Institution C. E., vol. vi, p. 256.

miles. Nevertheless, owing to the tendency of waves on that exposed coast to obstruct the mouth of the river with drift, the minimum depth in the outlet channel remained at about 4 feet below low water up to 1888. Since that time, however, dredging has been actively carried on, increasing the available depth to about 11 feet at low tide in 1892 (Plate 6, Fig. 7). Two converging breakwaters, moreover, are now in course of construction for sheltering the approach to the river; and the northern breakwater, which is approaching completion, will effectually prevent the drift from reaching the outlet of the river[1]; and it will be possible to dredge the outlet channel to any requisite depth within the sheltered area (Plate 6, Figs. 6, 8, and 9).

Breakwaters at Mouth of River Nervion. The breakwaters in progress across the small inlet on the southern shore of the Bay of Biscay, into which the river Nervion flows, are intended to form a harbour of refuge on that stormy coast, and to protect the entrance of the river which affords access to the port of Bilbao[2] (Plate 6, Figs. 12 and 15). They have been designed to extend out into depths of from 40 to 46 feet at low water, and will be considerably seawards of the site of the bar which formerly encumbered the mouth of the river, but has been removed by the training of the river (Plate 6, Fig. 13). These breakwaters, however, by protecting the beach near the mouth of the river from the violent waves which beat upon that coast, will prevent the bar from forming again, and will also protect the outlet from the easterly drift produced by north-westerly storms.

Remarks on Converging Breakwaters at River Mouths. Converging breakwaters are necessarily larger works than

[1] A plan and sections of the new breakwaters, and a longitudinal section of the present outlet channel, were sent me by Mr. H. H. Wake, engineer to the River Wear Commission

[2] 'Memoria que manifiesta el Estado y Progreso de las Obras de Mejora de la Ria de Bilbao, 1885–1889.' Evaristo de Churruca.

training jetties in prolongation of the river channel, and they become very costly when extended into deep water in exposed situations. This extension, though ensuring a deep outlet, and enabling any desired increase in depth to be provided by dredging within the sheltered area, is not expedient except at the approach to an important seaport, and where a refuge harbour is much needed, as at Tynemouth and the mouth of the Nervion. Breakwaters, however, merely converging towards the bar, may be essential for securing and maintaining a good outlet channel, especially where, as at the mouth of the Liffey, the tidal capacity of the river and the fresh-water discharge are small ; but the works, though carried out cheaply in Dublin Bay across a shallow foreshore, extended over a long period. The long breakwaters sheltering the Tees estuary were only rendered practicable by the refuse slag composing them being deposited free of cost.

The permanence of the tidal scour obtained by converging breakwaters depends on the absence of sediment in the inflowing flood tide, or in the maintenance of the tidal capacity of the area enclosed, by the periodical removal of any deposit that may take place. Owing to the great cost of large breakwaters, the system is only expedient where a small tidal capacity in the river, combined with a small fresh-water discharge, or the exposure of the mouth to heavy seas, renders training jetties inadequate for providing the requisite depth across the bar.

TRAINING JETTIES AT THE MOUTHS OF TIDAL RIVERS.

Tidal rivers may be trained, like non-tidal rivers, by cross jetties or longitudinal training banks, so as to obtain uniformity in depth by regulating the width of the channel (see Chapter III). The width, however, of the trained channel in tidal rivers must be gradually enlarged as it descends towards the sea to ensure the free admission of the flood tide, the rate of enlargement being increased on approaching the mouth of the river.   Thus the first improvement works on the Clyde consisted in the regulation of the river below Glasgow, first by cross jetties, and subsequently by longitudinal training walls, aided by the removal of hard shoals (Plate 8, Fig. 7). The Tees has been regulated in a similar manner, and also by the formation of straight cuts between Stockton and Middlesbrough (Plate 8, Fig. 4); and the Tyne has been improved by some straight cuts above Newcastle, by easing some bends below, and by shutting off some wide recesses near Jarrow (Plate 8, Fig. 1).   The Scheur branch of the Maas, moreover, has been systematically trained by longitudinal banks of fascine mattresses, where requisite for regulating the width, from Rotterdam to its outlet in the North Sea (Plate 6, Fig. 18); and the river Nervion has been similarly regulated by rubble training walls and quays, between Bilbao and its mouth (Plate 6, Fig. 12).   The mouth of a river, however, when opening directly on to the sea-coast, and exposed to the obstruction of sand or shingle drifting along the shore, generally presents the most difficult problem in any scheme of improvement.   Sometimes the action of the waves and currents upon the drift is so powerful that the outlet frequently changes its position, until it is fixed by jetties; and often works are required at the outlet to render a river accessible for vessels, or to enable ships to take advantage of the improvement in depth obtained by regulation works above.   Jetties training the outlet channel across the bar are simpler and more economical than the converging break-waters already referred to; and they often suffice, either to fix a wandering outlet, or to form a prolongation of the training works into deep water.

The jetties at the mouths of the Yare, the Adur, and the Adour, furnish examples of the fixing of shifting outlets (Plate 6, Figs. 4 and 10); whilst the jetties at the mouths of the Maas and the Nervion are instances of the prolongation of training works along rivers out to deep water, beyond the foreshore at the outlet (Plate 6, Figs. 12 and 16).

Jetties at Mouth of River Yare. Yarmouth Haven is now formed by the junction of the three rivers, the Yare, the Waveney, and the Bure, into a single channel above Great Yarmouth, which after running past the town, nearly parallel to the coast, for about 2¾ miles in a southerly direction, makes a sharp bend, and flows straight into the North Sea (Plate 6, Figs. 4 and 5). In Roman times, however, the river Bure appears to have had an independent outlet about 4 miles north of the town ; whilst the mouth of the Yare was only a mile south of the town. The Bure was eventually forced to join the Yare by the blocking up of its outlet by sand and shingle. Later on, however, the mouth of the Yare was driven about 4 miles further south by the drift occasioned by north-east storms ; and though a new channel was cut on several occasions between the fourteenth and sixteenth centuries, to bring the mouth of the river nearer Great Yarmouth, the new outlet was always sooner or later blocked up.

At last, in 1559, a cut was made at the site of the present outlet, and was fixed by two nearly parallel jetties across the foreshore, about 200 feet apart, which has since been maintained. The northern jetty has been extended in advance of the other, and has been made solid to protect the mouth from the drift coming along the coast from the north. The jetties maintain the outlet channel over the bar, which the drift tends to form in front of the mouth, by directing the fresh-water discharge and the tidal ebb across the foreshore. The rise of spring tides at that part of the coast is only 6 feet ; but owing to the flatness of the country through which the rivers flow, the tidal influence extends many miles inland ; and the tidal scour is promoted by the wide expanse of Breydon Water, into which the Yare and the Waveney flow above the town. Dredging also is employed to increase the tidal capacity of the river, and to deepen the channel. The minimum central depth in the jetty channel, at low-water spring tides, is now 10 feet, as compared with 7 feet in 1881; the minimum available depth over the bar is about 9 feet; and the available depth in the river up to the port is about the same, so that in calm weather the bar offers little impediment to the entrance of vessels.

Jetties at Mouth of River Adur. When the advance of the shore cut Old Shoreham off from the sea, the port was shifted to New Shoreham, situated on the river Adur which gives it access to the sea. The mouth, however, of the river was liable to be driven towards the east by the eastward drift of shingle along the coast; and in 1750 the mouth was $3\frac{1}{2}$ miles to the east of New Shoreham. Ten years later a new outlet was cut across the beach, a little more than a mile to the east of the town, and was protected at its extremity by two wooden piers; but when these piers were destroyed, the outlet travelled eastwards again, and at the close of the eighteenth century it was $3\frac{1}{4}$ miles east of the town. In 1819 the present entrance was cut through the beach, in almost the same position as the previous cut, and was again protected by more durable jetties, placing the entrance once more nearer New Shoreham.

As compared with the Yare, the Adur outlet possesses the advantages of a rise of tide of about 19 feet, and a somewhat more sheltered position; but, on the other hand, its drainage area is only 160 square miles, whilst that of the Yare is 1,228 square miles, the tidal flow up the Adur is much more restricted, and the foreshore on that part of the coast tends to advance.

Jetties at Mouth of River Adour. The mouth of the river Adour being situated near the south-eastern angle of the Bay of Biscay, where deep water approaches the coast, is exposed to the full force of the surf which beats upon that shore during north-westerly gales; and formerly its position was constantly shifted by the inroad of drift. Though the travel of the sand and shingle is mainly from north to

south, owing to the prevalence of north-westerly winds, the
mouth of the river, in 1450, had been driven 18 miles north
of its present outlet, which deviation, by obstructing its
outflow, threatened Bayonne with inundation during floods.
Accordingly, in 1579, a straight cut was formed to the
sea, and the old channel was barred, so that the river was
diverted into this new channel; and the depth over the bar
at the outlet was for a time improved. The drift from the
north, however, soon drove the channel towards the south,
and reduced its depth, till at last masonry jetties were constructed in 1732–41, which fixed the outlet in its present
position. The depth obtained at first by these works over
the bar was 20 feet at high water; but the bar soon rose
again further out. This result being attributed to the width
between the jetties, of 900 feet, being too great, it was reduced
eventually by low inner jetties to 600 feet; but the depth
over the bar was not thereby increased; and the outlet channel
shifted in position, exhibiting a tendency to travel south
again. Between 1808 and 1838, accordingly, the jetties
were prolonged, mainly on the south side, which prevented
the deviation of the channel to the south; and the width at
the outlet was reduced to 500 feet. These works, however,
whilst securing the stability of the outlet, only pushed the
bar further out, without reducing its height[1]. Prolongations
of the south jetty, with 656 feet of solid work and 656 feet
of open timber-work, on a rubble mound raised only to 6½
feet below the lowest low-water, and of the north jetty with
2,180 feet of open timber-work, on a low rubble mound, were
carried out in 1857–61. This open work was adopted for
the extensions to avoid a progression of the foreshore. The
timber-work having been attacked by the sea-worm, has been
replaced by iron cylinders, 6½ feet in diameter, with open
spaces, 10 feet wide, between them[2]; and the southern jetty

---

[1] 'Ports Maritimes de la France,' vol. vi, p. 931.
[2] Minutes of Proceedings Institution C. E., vol. lxx, plate 3, fig. 5.

is being prolonged, having been extended 200 feet since 1881 (Plate 6, Fig. 10). The intervals between the cylinders were intended to be temporarily closed by sliding panels whenever it might be expedient to concentrate the outgoing current; but this design has been abandoned, owing to difficulties in working, and its cost.

These extensions of the jetties have increased the depth over the bar from a minimum of 15¾ feet in 1857, to 17 feet in 1892, at high water of spring tides, the rise of tide at springs being 10½ feet (Plate 6, Fig. 11). The improvement in depth is, accordingly, very moderate as compared with the extent of the works; and both the foreshore and the bar have advanced seawards since the extension of the jetties; so that the stability of the mouth of the river is the main benefit realised by the jetties.

As the Adour drains only a small basin, and therefore has a small discharge except during floods, whilst the rise of tide at its mouth is fairly good, the narrowing of the channel, from about 1,000 feet near Bayonne to 550 feet at its outlet, was a mistake, considering that the maintenance of the outlet depends largely upon the tidal volume of the ebb, which has been seriously reduced by the contraction of the jetty channel, for the tidal range is decidedly less in the river than outside. The formation of an enlarging channel by diverging jetties, carried out beyond the bar, would have ensured the free admission of the flood tide; and the bar could have been lowered within the jetty channel by dredging. Solid jetties would have produced an advance of the foreshore; but probably the line of foreshore would have reached an equilibrium before compromising the outlet, with jetties extending beyond the bar. If adequate funds had been available, a still better outlet channel could have been obtained, on such a very exposed coast, by converging breakwaters extending into deep water, and dredging the outlet channel within the area thus sheltered.

Jetties at Mouth of River Maas. The general tendency of
the numerous channels of the Maas, which intersect South
Holland, is for the northern outlets to silt up, and the
southern outlets to become the only navigable channels,
owing partly to the large reclamations which have been carried
out along the northern channels, and partly to the larger rise
of tide to the south, which favours the maintenance of the
southern outlets. Accordingly, the growth of the bar at the
mouth of the most northern Scheur branch, which is the most
direct channel to Rotterdam, and at the mouth of the adjacent
Brielle branch, obliged sea-going vessels trading with Rotterdam to follow more southern, and very circuitous channels.
A canal was cut in 1829 through the island of Voorne, to
enable vessels entering by the deeper Hellevoetsluis entrance
to reach Rotterdam by the more direct Brielle branch, and
thereby avoid the Brielle bar. In course of time, however,
the increasing size and draught of vessels rendered the canal
inadequate for the requirements of navigation; and consequently it was decided in 1863 to form a direct channel
between Rotterdam and the sea, about 20 miles in length,
with a minimum depth of $21\frac{1}{2}$ feet at low water, by the
improvement of the Scheur branch[1].

The works which were carried out on the Scheur branch of
the Maas between 1863 and 1876, comprised the training
of the river in a gradually enlarging channel from Rotterdam
to the Hook of Holland; the formation of a direct outlet into
the North Sea across this neck of land, 3 miles in width; the
closing of the old circuitous outlet round the Hook; and
the training of the new outlet into deep water by slightly
diverging fascine-mattress jetties, 6,560 and 7,550 feet long
respectively, across the sandy foreshore (Plate 6, Figs. 16
and 19). The outer end of the north jetty was only raised
to half-tide level; but the south jetty was raised to ordinary

---

[1] 'Étude sur l'Effet des Marées dans la partie maritime des Fleuves.' P. Caland,
1861.

high-water level, in order that the flood tide coming from the south might enter the river straight up channel, and that the channel might be protected from the littoral drift from the south. The trained channel was made 738 feet wide at Krimpen above Rotterdam, increasing to 1,476 feet at Vlaardingen, and attaining a width of 2,953 feet at the extremities of the jetties. The new cut, three miles in length, was only partially excavated previously to the diversion of the river into it; and the widening and deepening of this channel was left to the scour of the fresh-water and tidal currents. The current, however, whilst deepening the narrow channel down to 40 feet below low water in some places, did not enlarge it much in width; and a great portion of the sand scoured out of the cut was deposited in the jetty channel, and arrested further scour. Accordingly, the minimum depth along the new waterway was only 12 feet at low water in 1877, after the completion of the works, the rise of tide at the mouth being only 6½ feet at springs, so that vessels drawing more than 18 feet could not get up to Rotterdam, only a slight improvement on the available depth of 16¾ feet by the Voorne Canal. Supplementary works were consequently commenced in 1882, for securing the required depth, which have now been completed.

The new works have been designed to remedy the defects in the original scheme, by enlarging the new cut to the requisite width, by excavation and dredging, to correspond with the channel above and below; removing the accumulations of sand in the jetty channel; reducing somewhat the enlargement of the trained channel; and preventing the loss of tidal water on the ebb, by checking its outflow into the Noordgeul channel leading to the Brielle outlet [1]. The trained channel has been reduced in width to 985 feet a little below Rotterdam, increasing fairly uniformly towards the sea, and

---

[1] 'Amélioration de la Voie Fluviale de Rotterdam à la Mer.' J. W. Welcker, Vᵐᵉ Congrès International de Navigation Intérieure, Paris, 1892.

through the enlarged cut, to 2,300 feet at the extremity of the jetty channel, which has been narrowed from its original width by a submerged fascine-mattress jetty, constructed about 650 feet to the north of the south jetty in continuation of the southern trained bank of the enlarged cut (Plate 6, Fig. 16). Extensive dredging works have also been carried out in the jetty channel, to remove the deposits and deepen the outlet channel, and also beyond the ends of the jetties to extend the navigable channel to deep water. As the Noordgeul channel, branching off from the Scheur branch nearly opposite Vlaardingen, could not be entirely closed, owing to its affording access to the Voorne Canal, its entrance width has been reduced from 1,066 feet to 230 feet, so that very little of the ebb current is now diverted from the Scheur branch. These works have increased the minimum navigable depth at the outlet to about 24½ feet at low water (Plate 6, Fig. 17); and whereas, in 1882, vessels could only reach Rotterdam with a maximum draught of 19½ feet, in 1888 vessels drawing 25 feet could effect the passage. Before any works were commenced, the circuitous journey between Rotterdam and the sea often occupied 18 hours; whereas now a vessel can traverse the 20 miles of channel between the sea and Rotterdam in 2 hours. The great improvement in the navigable access to Rotterdam, both in directness and depth, has transformed it as a seaport; and the increase in the navigable depth achieved by the recent supplementary works, has resulted in the trebling of the tonnage of the vessels trading with Rotterdam since 1879. A total sum of £2,965,000 has been expended, within the last thirty years, in the formation of the new channel.

Jetties at Mouth of River Nervion. The works of regulation and training which have been carried out, from 1873 to 1891, on the river Nervion between Bilbao and the mouth of the river on the south coast of the Bay of Biscay, a distance of nearly 8 miles, were designed to provide the developing port of Bilbao with a better navigable access to the sea[1]. The works, which have formed a channel enlarging in width from 210 feet at Bilbao to 525 feet at the extremity of the north-eastern jetty, resemble in principle the training works along the Scheur branch of the Maas ; but though on a much smaller scale, on account of the great difference in the width and length of the waterways, the training walls and jetty extension have been constructed with rubble-stone, concrete, and masonry, owing to the abundance of suitable materials in the neighbourhood (Plate 6, Figs. 12 and 14). A new cut was formed for the river, in place of a very tortuous channel, just above the confluence of the river Cadagua ; the channel has been rectified at other bends by rubble training walls and quays ; and excessive widths have been reduced by similar works. As the river only drains a small basin, its fresh-water discharge is small ; but the tidal rise at its mouth amounts to 14¾ feet at equinoctial springs, and is 9 feet at average tides.

The bar in front of the outlet, formed by littoral drift from the west under the influence of the prevalent north-westerly winds, and by wave-action on that exposed coast, has been lowered by the scour produced by the prolongation of the old Portugalete jetty, on the south-west side of the outlet, in a slightly concave line towards the channel, for a length of about 2,640 feet, extending out beyond the bar into a depth of 3 fathoms at the lowest low-water (Plate 6, Figs. 12 and 13). The jetty resembles a breakwater at its outer end (Plate 6, Fig. 14); but for the inner 1,870 feet, it is only solid up to high-water spring tides, and is surmounted by an open iron viaduct affording access between the old jetty and the new pierhead. The inclined direction of approach of the river to its outlet, causes the main outflowing current to keep

[1] 'Memoria que manifiesta el Estado y Progreso de las Obras de Mejora de la Ria de Bilbao,' 1879–1891 ; E. de Churruca ; and ' Anales de Obras Publicas,' vol. xi, 1883.

close along the south-western jetty. which is further assisted by the concave form of the jetty, so a prolongation of the old jetty on the opposite side of the channel was unnecessary. The scour and deepening produced by the extension of the jetty have been aided by dredging in the river. The improvement in the channel produced by the jetty, the training and regulation works, and dredging, is indicated by comparing the longitudinal sections of 1878 and 1889 (Plate 6, Fig. 13); for, with the exception of the river diversion which was commenced in 1873, the works were only begun in 1878. The minimum depth in the navigable channel of the river up to Bilbao, at the lowest low-water, has been increased from 5 feet in 1878, to about 12 feet in 1889. The depth over the crest of the bar has been increased within the same period from $3\frac{3}{4}$ feet to $15\frac{3}{4}$ feet at the lowest tides ; and though its crest, which has been pushed out seawards as far as the end of the new jetty, is still apparent, it presents no impediment to navigation, for it is about 3 feet lower than the bed of the river a little higher up. The exposure of the outlet, with deep water approaching the shore, and the liability of the mouth of the river to be obstructed by drift during storms, will be remedied by the shelter which the outlying breakwaters will eventually afford (Plate 6, Fig. 12). The expenditure on the training works and dredging since 1878 has amounted to £408,000.

Remarks on Works at River Outlets. With the exception of the Maas, the bars at the outlets of the rivers referred to in this chapter are of marine origin, and have been lowered by the scour across them, resulting from the works carried out at their mouths, aided by dredging. The scour, moreover, in the case of tidal rivers, has been augmented by the larger admission of tidal water up the regulated river, and the freer discharge of the ebb, or by the enclosure of a large area of tide-covered foreshore by converging breakwaters. The assistance of dredging is very useful in providing a greater depth than could have been attained by scour alone, and in maintaining the tidal capacity of a river when liable to be reduced by the deposit of silt; and frequently scour is able to maintain a depth which it could not have created unaided. Though the Maas brings down from inland most of the material which encumbers its outlets, the tendency of the sea to encroach along that coast prevents a progression of the foreshore; and the alluvium is carried northward by the flood tide.

The construction of the jetties depends upon the available materials. Where stone is easily procured, as at Tynemouth, Dublin Bay, and the mouth of the Nervion, the jetties are naturally formed with it. Where, however, as at the mouth of the Maas, and at the delta of the Mississippi, there is no stone in the vicinity, but osiers are cultivated in abundance, fascine work constitutes the most suitable method of construction, especially on a soft sandy or silty foundation (Plate 6, Fig. 19). The employment of fascine work for structures in the sea, first adopted at the mouth of the Maas, might have appeared rather too bold an expedient; but it has been justified by experience, for the fragile material is soon protected from decay by a coating of sand or silt, whilst the yielding nature of the fascine work enables it to withstand the shocks of the waves without injury. Fascine mattresses have since been resorted to in the more extensive jetties for improving the entrances to Charleston and Galveston harbours[1], as well as for the South Pass jetties at the mouth of the Mississippi (Plate 5, Fig. 8).

The mouths of rivers are not generally altered in position, except when a shifting outlet has to be fixed by jetties; but in the case of the Maas, it was possible to obtain a more direct outlet, and better depths inshore than at the natural

---

[1] 'Harbours and Docks.' L. F. Vernon-Harcourt, plate 2, figs. 1, 2, 12, and 13.

outlet, by a cut of moderate length across a low neck of land.

The great advantages for navigation that may be secured by jetty works at the mouth of a river, have been amply demonstrated by the instances described, especially if supplemented by training works and dredging carried up to the port for which improved access is required.

CHAPTER X.

TIDAL FLOW IN RIVERS; AND FORMS OF ESTUARIES.

Tidal Flow in a Creek; its effects. Tidal Flow and Fresh-water Discharge in a River. Alluvium in a Tidal River; its origin; instances. Influence of Fresh-water Discharge; advantages of large discharge; examples. Influence of Tidal Flow; importance of tidal volume; contrast of state of tidal and tideless rivers. Influences of Flood and Ebb Tides; periods of greatest velocity; divergence in directions. Conditions affecting Tidal Flow; large tidal rise; small fall of river; examples. High-water line; general uniformity; instances of rise; irregularities. Lowering of Low-water Line; advantages; indication of improvement. Progress of Tide up a River; indicated by simultaneous tidal curves; acceleration of flood tide. Bore, cause, instances. Obstructions to Tidal Flow, effects, disadvantages. Effect of Tidal Flow on Outlet. Forms of Channels and Estuaries; best theoretical form; varieties in forms of estuaries. General Principles relating to Tidal Rivers.

THE physical conditions of tidal rivers are more varied than those of the non-tidal portions of rivers, or of rivers flowing into tideless seas, owing to the differences which exist in the tidal rise at different parts of the sea-coast, in the distance to which the tidal influence extends up a river, and in the tidal capacity of a river.

Influence of Tidal Flow in a Creek. When the tide flows up a sheltered creek, into which there is no discharge of fresh water from the land, if the flood tide enters charged with sand or silt, the creek will gradually silt up; for, though the ebb tide carries out again a large portion of the material brought in by the flood, a portion of the sediment, deposited during slack water at the turn of the flood tide, is left behind. The flood tide, which is impelled up a creek by the difference in head between the water in the sea and in the creek, has its force diminished by the inert mass of water previously in the creek, and by friction against the sides of the creek, which also reduce the force when reversed on the ebb.   Accordingly, the force of the ebb in the creek is less than the force of the flood by the sum of the retarding forces acting during both the flood and ebb tides.   The silting up is also promoted by the sediment, when once deposited, being less easily removed than when held in suspension.

Tidal Flow, and Fresh-water Discharge in a River.   The tidal flow in a river differs in two respects from the tidal flow in a creek.   In the first place, a river affords a much longer distance for the run of the flood tide than a mere creek, and therefore the tidal currents are stronger and more continuous, presenting consequently less opportunities for the deposit of sediment transported by the water.   Secondly, the ebb tide is reinforced by the fresh-water discharge of the river, which is penned back during the flood tide ; so that the whole volume of fresh water brought down by a river during a period of rather over twelve hours, has to flow away during the ebb.   Accordingly, a tidal river is much less subject to accretion than a simple creek.   The influence, indeed, of the fresh-water discharge of a tidal river in maintaining the depth of the river, is manifested by the bars at the mouths of rivers being frequently raised when the discharge is small, and being lowered again on the descent of floods.   When the tidal influence extends some distance up a river, the fresh-water discharge during the tidal period does not reach the sea in a single ebb tide, but it proceeds down the river in successive stages, and is equally efficacious in reinforcing the ebb.   The tidal oscillations of the flood and ebb maintain the water in almost constant motion, which hinders deposit ; whilst the fresh-water discharge counterbalances the preponderating energy of the flood tide.   A kind of equilibrium is, accordingly, generally maintained in tidal rivers, until modifications are introduced into the existing conditions, either by engineering works, or by the slow changes produced by the irresistible operations of nature. If a state of equilibrium was not attained, a river would either continue deepening its channel till its scouring power was sufficiently reduced, by the increase in the size of its bed, to produce no further effect, or its channel would gradually silt up till the concentrated current was able to maintain the diminished channel.

The fresh-water discharge of a river is a definite volume, proportionate to the area of the basin and the rainfall over that area, which varies according to the seasons, and fluctuates from year to year with the rainfall, but approximates to a certain average when measured during a series of years. This volume of water has to be discharged by the river ; it cannot be augmented; but it can be diminished in amount,either by abstraction for water supply, or by diverting it into a new channel. The tidal volume, on the contrary, is practically unlimited ; it can be increased by facilitating its influx into a river, and by removing obstacles to its upward flow ; or it can be reduced to any extent, by restricting the width of the entrance to a river, and by barring its progress above. Accordingly, a tidal river possesses capabilities for improvement or injury, which are absent in a tideless river.

Alluvium in a Tidal River. Alluvium may be introduced into a tidal river from inland by the fresh-water discharge, as in the case of tideless rivers ; or it may be brought in by the flood tide, having been stirred up by waves from shoals, or eroded from the neighbouring sea-coast. In the former case, the river is liable to deposit its burden in the upper tidal portion, when arrested in its flow by the flood tide. In the latter case, the incoming flood tide tends to deposit the heavier sediment in any sheltered places near the mouth of the river, out of the main run of the tide, and the lighter alluvium higher up during slack water at the turn of the tide, particularly when the fresh-water discharge is small. The greater part of the alluvium from inland is brought down by floods, which convey it further down the river, before it is arrested by the flood tide, in proportion as the flood discharge overpowers the enfeebled tidal flow near the tidal limit; and it is thus brought under the influences of a greater exposure and stronger tidal currents, which prevent its being readily deposited. The flood tide at springs, being larger in volume and more rapid in flow than at neaps, brings in a greater quantity of material; but the ebb tide at springs, being for the same reasons more powerful, counteracts this tendency to accretion, except in sheltered creeks and near the limit of the tidal flow. The flood tide comes in specially loaded with detritus after storms. Accordingly, a river is most exposed to the accumulation of sediment in its upper tidal portion when the fresh-water discharge is small, and near its outlet after storms. Nevertheless, the occurrence of slack tide at different places all along the tidal portion successively, obviates the continuous accumulation of deposit in any particular spot, such as takes place beyond the outlets of silt-bearing tideless rivers; whilst the almost constant tidal oscillations are unfavourable to silting-up.

The alluvium in most tidal rivers is of mixed origin, being partly derived from inland, and partly from the sea, varying in proportion according to the physical conditions of the river, and of its adjacent seaboard. Thus the alluvium of the Ganges comes wholly from inland; and the fresh-water discharge in this case is so large, and is so densely charged with alluvium, that, in spite of the rise of tide of about 12 feet at the mouths of the Ganges in the Bay of Bengal, the river has formed a delta like sediment-bearing rivers flowing into tideless seas. The alluvium in the Seine estuary, on the contrary, is almost entirely of marine origin, for the fresh-water discharge of the Seine is remarkably free from sediment; whereas the flood tide sweeping along the coast of Calvados before entering the Seine, brings large quantities of detritus into the estuary from the eroded cliffs. This, moreover, is indicated by the good depth which has always existed in the upper half of the tidal Seine; whilst the lower half was encumbered by sands (Plate 9, Figs. 1 and 2). The alluvium in the Loire estuary affords an example of mixed origin; for ·this torrential river brings down large quantities of material in flood-time, which have accumulated in sheltered parts of its upper estuary; whilst the sands of its outer estuary are of marine origin (Plate 9, Fig. 4). Those rivers are naturally most easily maintained and improved, which are least exposed to the introduction of alluvium, either from inland or from the sea.

Influence of Fresh-water Discharge. It has been already pointed out that the fresh-water discharge of a tidal river reinforces the ebb tide, and consequently is favourable to the maintenance of a tidal river. Provided the fresh-water discharge is not densely charged with alluvium, and is not so large as to overpower the flood tide and thus prevent the tidal oscillations, making the river approximate to the condition of a tideless river, the value of the discharge in maintaining a tidal estuary is proportionate to its volume. Thus the tidal Seine, with an average fresh-water discharge into its estuary of 28,000,000 cubic yards of fairly clear water every tide, is in this respect in a much more favourable position than the Mersey estuary, which receives a fresh-water discharge averaging only about 2,500,000 cubic yards per tide of more turbid water. Though a large fresh-water flow reduces proportionately the volume of tidal water that can enter a tidal river, this loss of tidal flow does not signify, since the volume of the ebb is maintained, unless the fresh water brings down more alluvium than can readily be dispersed, in which case the zone of deposit is unfavourably concentrated by the restriction of the limits of the tidal oscillations.

The amount of alluvium, however, brought down by a river can only be very slightly modified by protecting its banks from erosion; and therefore the scope for the removal of defects resulting from the condition of the fresh-water discharge of a river, is very limited compared with the alterations that can be effected in the tidal flow.

A small fresh-water discharge has the disadvantage of rendering the transition from the tidal to the non-tidal portion of a river very abrupt, where the river flows into a large estuary, and the tidal rise at its mouth is considerable. Thus the river Mersey, which affords access to the largest class of vessels up to Liverpool, becomes unsuitable for ordinary inland navigation before its tidal limit is reached, about 46 miles from the bar at its outlet. The Thames also, which is the waterway, below London Bridge, for the largest port in the world, can only give accommodation to small river craft near its tidal limit. A large fresh-water discharge, on the contrary, enables vessels of good draught to penetrate long distances inland. Thus the Rhine, with a basin about fifteen times that of the Thames, provides an open navigable waterway far inland up to the flourishing ports of Mainz and Mannheim; and Rouen, the fifth seaport of France, is 75 miles from the mouth of the Seine.

Influence of Tidal Flow in Rivers. The most striking characteristic of the influence of tidal flow is the navigable importance with which it invests rivers with a small fresh-water discharge, and which without its aid would be of no value to ocean navigation. Great Britain, indeed, owes her commercial pre-eminence to the tidal wave which travels round her coasts; for her rivers, with their small drainage areas, could never, without tidal influences, have provided access for the commerce of the world. The channels of non-tidal rivers are merely proportional to the average discharge; and in dry weather the flow is liable to become so small as to afford a very inadequate depth for navigation. As pointed out in Chapter III, this variation in discharge renders non-tidal rivers unsuitable for continuous navigation throughout the year, unless, from the great size of their basins, or from differences · in the period of the low-water flow of their tributaries, their discharge never falls very low. Tidal flow, however, under favourable conditions, very largely increases the volume of water passing through the channel, and also furnishes the additional supply requisite for filling an estuary when the fresh - water discharge is reduced. Moreover, the volume of tidal water entering and leaving a tidal river at each tide, though varying somewhat between springs and neaps, fluctuates within far narrower limits than the fresh-water discharge. Accordingly, a tidal river, though rising and falling with every tide, is much more regular in the volume of its flow than a non-tidal river; and it is able to maintain an outlet channel enormously larger than could be maintained by its fresh-water discharge. The Mersey furnishes an instance of extreme disparity between tidal and fresh-water flow; for at a high spring tide, the tidal volume entering and leaving the Mersey estuary is seven hundred and ten times the average volume of fresh water discharged into the estuary each tide.

The value of tidal flow in a river is forcibly illustrated by contrasting the size of the Thames above Teddington and at Gravesend, and of the Mersey above Woolston weir and at Liverpool. The influence also of tidal flow in placing small rivers on a par, as regards navigability at their outlets, with some of the largest rivers flowing into tideless seas, is readily manifested by a comparison of their conditions. Thus the Thames, having a basin only $\frac{1}{62}$ that of the Danube, affords superior facilities for navigation at high water between the sea and London than the Sulina mouth; the Mersey, with a basin only $\frac{1}{722}$ that of the Mississippi, is equally accessible at high tide up to Liverpool as the South Pass; the Ribble, with a basin $\frac{1}{58}$ that of the Rhone, is of more use for navigation up to Preston than the Rhone outlet; and the Usk, with a basin only $\frac{1}{888}$ that of the Volga, has

nearly double the depth in its navigable channel at high
water of fair neap tides up to Newport, which it is hoped
may be obtained by dredging at the most favourable Volga
outlet, and about three times the present available depth
in that river from the Caspian Sea up to Astrachan.

Influences of Flood and Ebb Tides. The flood tide running
up a river, flows with a greater velocity, and has a more
powerful scouring effect during the earlier half of its flow,
when the channel is more restricted, and before the banks
in an estuary are covered. The greater density of sea-water
augments the scour of the flood tide, by causing the incoming
current to pass along the bottom under the lighter water
of the ebb, which is more or less mixed with fresh water.
The flood tide having thus at first to overcome the friction
against the bottom, and the opposing current of the last of
the ebb, does not attain its maximum velocity till a little
time after the turn of the tide. The scour of the ebb is
greatest during the latter half of its flow, when it runs in
a more restricted channel, and with a greater inclination
owing to the fall of the tide outside; and the ebbing current
attains its greatest velocity at about half ebb, when the banks
in an estuary begin to uncover, and the waters begin to
converge towards the low-water channel.

A river is in the best condition when the flood and ebb tides
follow the same course, and thus both assist in maintaining
the depth; but this is not generally the case in a wide
channel, or in an estuary, for the flood tide adopts as straight
a course as possible, whereas the ebb tide runs close along the
concave banks of a river, shifting from side to side, like the main
current in a non-tidal winding river. The flood and ebb currents
are, accordingly, conflicting in their actions, the one tending to
scour out where the other tends to deposit. The ebb current
forms the fairly continuous winding low-water channel found
in most wide tidal rivers and estuaries; whilst the flood tide
forms the straight blind channels on the opposite side to the

ebb channel, which the action of the ebb prevents it from completing. These blind channels are clearly defined in the low-water charts of the estuaries of the Humber, Mersey, Clyde, Seine, Weser, Loire, and Ribble, and in the plan of the river Usk (Plates 7, 8, and 9). In the narrower upper portions of tidal rivers, the flood and ebb are necessarily confined to the same channel; but even under these conditions, the reversal of the direction of the current prevents the action of the flood and ebb in a winding channel being quite uniform. The preponderance, however, of the ebb tide, due to the fresh-water discharge, increases with the distance from the mouth, owing to the gradual diminution of the tidal influx in ascending a river.

The introduction of the early flood tide under the ebb current, and the divergence in the directions of the flood and ebb, enable the ebb current to continue flowing down during the first of the flood; and therefore the sea water gets higher up a river, and the fresh water lower down in each tide, than if the flood and ebb currents were directly opposed. Accordingly, there is a less direct backing up of the fresh water, and a greater intermixture of salt and fresh waters, with their burdens of alluvium, than might have been anticipated, which is further aided by eddies resulting from irregularities in the banks and bed.

Conditions affecting Tidal Flow. A large rise of tide at the mouth of a river affords a good navigable depth at high water, facilitates the introduction of the flood tide, creates energetic flood and ebb currents, and ensures the propagation of the tidal influence some distance inland; and it is consequently favourable to the maintenance of a tidal river. The Dee, the Mersey, the Ribble, the Severn, and the Usk are peculiarly favoured in this respect (Plate 7, Fig. 12; Plate 8, Fig. 10, and Plate 9, Figs. 12 and 14); whereas the Nervion, the Maas, the Loire, the Weser, and the Clyde are less fortunately circumstanced (Plate 6, Figs. 13 and 17; Plate 8, Fig. 8,

and Plate 9, Figs. 5 and 9). The distance up a river to which
the tide extends is also largely affected by the rise inland of
the river-bed. When the inclination of the river-bed is very
small, owing to the flatness of the country through which the
river flows, as in the case of the Maas, the Scheldt, the Seine,
and the Thames, the tide penetrates a considerable distance up
the river. Thus the tide flows up to Ghent on the Scheldt,
110 miles from the sea ; to Martot on the Seine, 90 miles from
Havre ; and to Teddington on the Thames, 67 miles from the
Nore. The length of the tidal Scheldt is due to the flatness
of the low-lying lands of Flanders through which it flows, for
the rise of tide at its mouth is less than at the mouth of the
Seine and of the Thames. The length of the tidal Mersey,
on the other hand, is less than that of the tidal Weser, owing
to the rise of the bed of the river above Eastham, though the
average rise of tide at the mouth of the Mersey is about twice
the rise at the mouth of the Weser (Plate 7, Figs. 3, 5, and 12,
and Plate 9, Fig. 9). The extent of the tidal influence,
accordingly, varies in proportion to the tidal rise and inversely
as the rise of the river-bed ; and the position of the tidal limit
changes with the differences in rise between springs and neaps,
and is, moreover, affected by the volume of the fresh-water
discharge, which in flood-time pushes the tidal limit seawards.

High-Water Line in a Tidal River. The high-water line
along a river is generally fairly level, and is not susceptible of
much alteration by improvement works carried out in a river.
When, however, an estuary faces the direction of the travel
of the tidal wave and is funnel-shaped, the level of high water
is raised in ascending a river by the concentration of the tidal
wave. The raising of the tidal wave in flowing up the British
Channel is a well-known instance of this influence, causing the
range of tide at the mouth of the Severn to become about
double the range in the open sea outside. The rise of the
high-water line is also noticeable in the Seine and the Weser,
resulting from the above causes (Plate 7, Figs. 1 to 4); but

the rise in the case of the Mersey, though due to similar causes at Liverpool, must be attributed to the rapid rise in the river-bed in the upper part of the estuary from the accumulation of sand (Plate 7, Figs. 5, 6, and 12). When the outlet of a river is restricted in width, without a corresponding deepening in the narrow part and shoaling in the wider channel above, or when the channel of a river is abruptly narrowed, the ingress of the flood tide is checked, and there is a fall in the high-water line. Thus the contracted outlet of the Adour reduces the rise of tide in the river (Plate 6, Figs. 10 and 11); and the narrowness of the trained channel of the Seine causes a fall in the high-water line above Quillebeuf (Plate 7, Fig. 1, and Plate 9, Fig. 1). Irregularities in the channel also produce irregularities in the tidal flow. A rise in the high-water line inland affords a river a surplus tidal capacity beyond the normal high-water level at sea, and an extended tidal influence; whilst a depression of the high-water line causes a loss of tidal capacity, and a reduction in tidal influence.

Lowering of the Low-Water Line   The low-water line depends upon the depth and regularity of the channel; and it can be lowered considerably by the deepening of the channel, and the removal of shoals which impede the efflux of the ebb. The promotion of the outflow of the ebb increases its scouring efficiency, by causing a larger body of water to flow down at each tide, and augmenting its fall near low water. The more thorough efflux of the ebb, moreover, brings more rapidly seawards any alluvium carried down by the fresh water; and by emptying the tidal channel more completely, the tidal capacity of the river is increased, and a greater volume of tidal water flows and ebbs through the outlet at each tide, promoting the increase of its depth. Provided the width of the entrance to the river is adequate, the increased tidal capacity does not affect the high-water line; for whilst a larger volume of tidal water has to enter at each tide, its influx is facilitated by the deepened channel. The lowering of the low-water line by dredging and training works, is indicated on the longitudinal sections of the Wear, the Tyne, the Tees, the Clyde, the Seine, the Weser, and the Ribble (Plate 6, Fig. 7, and Plates 8 and 9). The increase in tidal capacity, however, thus indicated, only applies to the navigable channel; and though correctly exhibiting an improved tidal flow in the channel itself, and in narrow rivers like the Tyne, and the upper portions of the Clyde and the Weser, it proves nothing as to the actual influence of training works in wide estuaries, like the Tees, the Seine, and the Ribble, on the tidal capacity as a whole. Nevertheless, with this reservation, the lowering of the low-water line indicates a distinct improvement in the condition of a tidal river, especially in its navigable channel; for the greater freedom of discharge thereby exhibited, manifests a deepening or rectification of the main channel.

Progress of the Tide up a River. The high- and low-water lines on the longitudinal sections of rivers are merely convenient representations of the highest and lowest levels of the water at various places on the river (Plates 6 to 9), for high and low water never occur all along a river at the same time. It is, indeed, sometimes high water at one part of a river when it is low water at another; and the actual water-lines, and the progress of the tide up a river are shown by simultaneous tidal curves, obtained by recording the height of the tide at a series of tide gauges along a river simultaneously (Plate 7, Figs. 1, 3, and 5). The tidal oscillations at each station are obtained continuously by recording the changes in level of a float, by means of a pencil marking a line on a sheet of paper wound round a revolving cylinder, thereby producing a tidal diagram of the heights of the river at the periods indicated (Plate 7, Figs. 2, 4, and 6). It will be noticed that on ascending a river the duration of the flood tide constantly decreases, also that the rise of the flood tide is very abrupt in some cases, owing to the opposition offered to its progress by sandbanks, and that the flood tide would go further up the Seine if not arrested by

a weir at Martot.   The duration of high water for $2\frac{1}{2}$ hours
at the mouth of the Seine (Plate 7, Fig. 2) is due to the
entrance of a second tidal wave, travelling more slowly along
the coast of Normandy, after the primary tidal wave, coming
into the estuary from deep water in the English Channel, has
passed.   This is a purely local condition, due to the configuration of the coast, similar to the double tide at Southampton,
but it materially facilitates the access of vessels to the port of
Havre.   It must, however, somewhat conduce to the deposit
of material in the estuary, owing to the long period of fairly
slack water, only partially interrupted by a reverse current
flowing out of the estuary near Havre.   The flood tide has
a shorter duration than the ebb, even near the outlet of a river;
and its period, as well as the extent of its influence, is reduced
by a large fresh-water discharge.

Owing to the differences in the time of high water in a river,
a tidal river is never completely filled, as, indeed, is evident
from an inspection of the simultaneous tidal curves of the
Seine, the Mersey, and the Weser (Plate 7, Figs. 1, 3, and 5);
and allowance must therefore be made for this deficiency in
estimating the tidal capacity of a river.   It is possible, accordingly, to increase the tidal volume in a river by facilitating the
influx of the flood tide by the removal of shoals or other
obstructions, and thereby making the tide rise higher in the
inner parts of a river before the turn of the tide at the mouth.
This result is indicated by the occurrence of high water at
places along a river at shorter intervals after high water at the
mouth than previously, such, for instance, as has been effected
on the Tyne by dredging, where the interval between high
water at Newcastle and the mouth has been reduced from
one hour to twelve minutes.   An advance, on the contrary,
of the time of high water at the mouth of an estuary, close
to the open sea, proves that the estuary becomes full earlier
than formerly, and therefore that its tidal capacity has been
reduced, which is exemplified by the case of the Seine, where

the training works in the estuary have been followed by very extensive accretions, and the time of high water at Havre has consequently been advanced 38 minutes.

Bore on Rivers. When the progress of the flood tide up a river is impeded by extensive shoals, the tide has to rise at the mouth till it acquires a sufficient head to overcome the friction caused by the obstruction ; and then it forces its way up as a rapidly-travelling crested wave, producing, as it passes along, an instantaneous reversal of the current, and a sudden rise of the water in the river. This phenomenon, which is termed a *bore*, is developed mainly at spring tides, and is increased by wind and by a funnel-shaped channel ; it is a well-known occurrence on the Severn, the Seine, the Hooghly, and the Amazon; and it was observed by Capt. Moore, R.N., on the Tsien-Tang Kiang[1], in China, in 1888. The bore on the Severn is about 6 feet high near Newnham ; and on the Hooghly, it attains a height of over 7 feet. On the Amazon, it appears like two or three waves in succession, from 12 to 15 feet high ; and on the Tsien-Tang Kiang, its maximum height is said to vary between 10 and 15 feet. The bore on the Seine reaches its maximum, of about 6 feet, at Caudebec, as, indeed, is evident from the simultaneous flood-tide curve between that place and Honfleur (Plate 7, Fig. 1), which represents to a distorted scale the true form of the bore, with an almost vertical breaking face in front, and a slight rise, instead of a hollow like a wave, behind.

Effect of Obstructions placed across Tidal Rivers. It has sometimes been assumed that, as the tide when flowing into bays or creeks tends to fill them up gradually with sand or silt, therefore the tidal flow and ebb in rivers is injurious, and that rivers would be in a better condition if the tide was excluded from them, as the sedimentary matter brought in by the flood tide would be thus shut out, and not be deposited in

---

[1] 'Report on the Bore of the Tsien-Tang Kiang,' Commander Moore, R.N., London, 1888; and a Further Report in 1892.

the river channel.   This opinion, however, is not borne out by
experience.   For instance, the harbour of Rye, at the mouth
of the river Rother, was injured by reclamations, and by the
erection of Scott's sluice, which, stopping the flow of the tide
up and down the river, caused a rapid accumulation of silt in
the channel below the sluice; and the subsequent destruction
of the sluice produced an improvement in the outlet.   It is
related by Mr. Cordier that the harbours of Dunkirk and
Gravelines suffered injury from a similar cause[1].   Mr. Bouniceau
describes the deterioration of the channel of the river Vire, in
Vays Bay, by the erection of tidal gates across the river near
Isigny, and its restoration to its original condition, on giving
free admission again to the tide by the removal of the gates[2].
Denver Sluice, placed across the Great Ouse, soon occasioned
a silting up of the river below; and the Grand Sluice across
the Witham, just above Boston, restricting the tidal flow
which formerly extended up to Lincoln, thirty miles higher
up the river, seriously injured Boston as a port.

A deterioration of the channel below is, indeed, the inevitable result of the stoppage of the tidal flow, when the flood
tide is charged with silt.   When the tide flows up a river, it
possesses a definite velocity which enables it to hold in suspension the sand and silt which it carries up with it; but on
encountering any barrier which prevents it from flowing
further up, it simply rises vertically against the barrier, and
deposits, during the long period of almost absolute slack
water thus produced, nearly all the sand and silt which it
brought up.   Thus before the excavation of the new outlet
channel for the Witham, the silt was sometimes heaped up
as high as ten feet against the gates of the Grand Sluice
at Boston in the summer, and was only removed by the
winter floods.   Moreover, the effect of such a barrier is not
confined to the portion of the river in its immediate vicinity;

---

[1] 'De la navigation intérieure du Département du Nord,' J. Cordier, p. 105.
[2] 'Étude sur la navigation des rivières à marées,' P. Bouniceau, p. 62.

for though the tidal flow is only absolutely arrested at the barrier, it is affected for a considerable distance down a river, in proportion to the proximity of the barrier and the burden of sediment brought up by the flood tide. For instance, the limit of the tidal Yorkshire Ouse has been restricted by a lock and weir at Naburn, to increase the navigable depth between that place and York; whilst the flood tide brings up large quantities of silt with which the waters of the Humber are always densely charged. In dry weather, when the fresh-water discharge of the Ouse is very small, this silt deposits to such an extent during slack tide in the upper part of the river, where the tidal scour is necessarily feeble, that towards the close of a dry summer, the bed of the river is 3 or 4 feet higher than in the spring, for 12 miles below the lock, and is only restored to its normal condition by the winter floods.

The tidal flow on the Thames was greatly impeded by the obstruction presented by Old London Bridge, which formed a sort of weir across the river, with its numerous piers resting upon wide foundations protected by piles, so that spring tides rose somewhat higher, and fell about 4 feet lower below the bridge than above; and the checking of the flood tide caused an accumulation of silt below the bridge. The removal of Old London Bridge, on the completion of the new bridge in 1831, and the substitution of bridges with fewer piers and larger waterways for the old bridges at Blackfriars and Westminster, aided by some dredging and the constant removal of gravel and sand from the bed of the river for building operations, have very materially improved the tidal condition of the river, so that the range of spring tides has been increased by $5\frac{3}{4}$ feet at London Bridge, and the low-water line has been lowered $4\frac{1}{2}$ feet at Teddington; whilst the silt has been scoured away below London Bridge.

Experience, accordingly, demonstrates that obstructions to the tidal flow are injurious to the depth of a river for some distance below them ; and the value of the flux and reflux of the tide in maintaining a channel, is manifested by the deterioration which results from a diminution in the tidal volume passing up and down a river, and the improvement which follows on its restoration.

Effect of Tidal Flow on the Outlet of a River. It has been seen that the outlet of a river is the most critical point of the whole channel, and that, owing to the diminution in velocity which the current generally experiences at that part, it is the place where a shoal, or bar, is most likely to form. Slack water is the period when deposit mainly occurs, and hence it is of great importance to reduce this period as much as possible. It is evident that in a simple creek, the longer and larger the creek is inside, the better will its outlet be maintained ; for the period of slack water will be reduced by the tide taking a longer period to flow up and down the creek, and a stronger current will be produced at the outlet. The deterioration of the outlet channels of the harbours of Calais and Ostend, as the shallow tide-covered areas within the harbours were reclaimed, furnishes examples of the great injury that may be effected in an outlet by the simple reduction of the tidal volume passing and repassing through the entrance. As regards, therefore, the outlet channel, it will be better maintained in proportion to the tidal capacity of the river above it.

Forms of Tidal Channels and Estuaries. As it is impossible to prevent the introduction of alluvium into a tidal river, either from inland by the fresh-water discharge, or from the sea-coast by the flood tide, the object to be aimed at is to prevent its accumulation in the channel of a river, and a consequent reduction in the navigable depth. Deposit occurs when any considerable reduction takes place in the velocity of a current charged with alluvium ; and therefore it is important that uniformity in the rate of the tidal ebb and flow should be maintained as far as possible, combined with the free admission of the flood tide. To comply, accordingly, with theoretical requirements, the mouth of a river should be large enough to admit freely at every tide the full volume of tidal water that the river is capable of receiving; and the sectional area of the river should gradually diminish inland, in proportion as the tidal volume required for filling the receptacle above it becomes less in ascending the river. A contraction at the mouth of an estuary, by checking the influx of the flood tide, and also increasing the velocity of the current through the narrow outlet, reduces the volume of tidal water entering and leaving the river, shortens the extent of the tidal influence up the river, and promotes deposit by the variation in the rate of flow. A tidal river, therefore, gradually enlarging in width as it approaches the sea, so as to secure the free admission of the flood tide and uniformity of flow, affords the best prospects of maintenance and uniformity in depth.

The actual forms of estuaries are very varied, depending upon the configuration of the coast, the physical conditions of the surrounding country, and the nature of the strata in which the estuary is situated; and they are all more or less encumbered with sandbanks and shoals, between which the low-water channel of the river meanders, and which are mostly covered at high tide (Plates 7 to 9). Some rivers enlarge fairly regularly on approaching the sea, as for instance the Thames and the Clyde (Plate 7, Fig. 7, and Plate 8, Fig. 7); whilst others widen out very rapidly, like the Ribble and the Weser, or emerge abruptly into a large estuary like the Tees, the Seine, and the Dee (Plate 8, Fig. 4, and Plate 9). Many estuaries exhibit irregularities in width, such as the Humber (Plate 7, Fig. 9), the Scheldt, and the Shannon[1]; whilst others

---

[1] Plans and longitudinal sections of these and other estuaries are given in the plates of my paper on ' The Physical Conditions affecting Tidal Rivers; and the Principles applicable to their Improvement.' Manchester Inland Navigation Congress, 1890.

are contracted in width at, or near their mouth, as for example the Mersey (Plate 7, Fig. 11) the Loire (Plate 9, Fig. 4), the Tay, and the Gironde. Some tidal rivers, on the contrary, enter the sea direct without passing through any estuary, like the Tyne (Plate 8, Fig 1), the Wear, the Nervion, and the Maas (Plate 6); but the outlets of such rivers have generally to be protected by jetties or breakwaters. The estuaries which are most regular in width, with a trumpet-shaped outlet, are also most uniform in depth, like the Thames and the Usk (Plate 7, Figs. 7 and 8, and Plate 8, Figs. 9 and 10); whilst estuaries which present great irregularities in width, with a narrow outlet, exhibit also great variations in depth, of which the Mersey and the Loire furnish examples. Gradual accretion in recesses often reduces the irregularities of an estuary in respect of the low-water channel, making it adopt a more trumpet-shaped form and become more regular in depth, as indicated in the estuaries of the Humber and the Tay. These deposits, which are the natural consequences of irregularity of flow due to variations in width, though reducing the tidal flow and ebb through the outlet, produce some compensation in the channel above by facilitating the influx higher up of the tidal water which previously spread over the recess, and are of no importance in estuaries, where, as in the case of the Humber and the Tay, the depth at the mouth and out to deep water is ample. In an estuary, however, like the Mersey, where the depth over the bar is already inadequate, and where the existing depth of the outlet channel depends upon the tidal capacity of the upper estuary, which acts like a sluicing basin, any material diminution in this tidal capacity would be prejudicial to the depth over the bar, and should be carefully avoided as far as possible.

General Principles relating to Tidal Rivers. The value of the tidal ebb and flow in rivers has been demonstrated by the commercial importance bestowed by the tide on rivers possessing a small fresh-water discharge, and the contrast which many of them present at their outlets to tideless rivers of great size. The importance also of promoting the influx of the tide up a river to the fullest practicable extent, has been proved by the injuries produced in the navigable condition of tidal rivers by restricting the tidal influence. Accordingly, the primary objects in all schemes of tidal river improvement, should be the admission of the flood tide as freely as possible into a river, and the removal of all obstructions to its progress up to the furthest practicable limit. Not only is the tidal volume passing up and down the channel thereby increased, leading to an enlargement of the channel; but the period of slack tide is thus reduced to a minimum, and the tendency to deposit greatly diminished.

As the fresh-water discharge gives a preponderating influence to the ebb tide, and thus materially aids in the removal of alluvium seawards and in the maintenance of the channel, it should not be diverted from its beneficial action at the head of a tidal river, except for the necessary paramount claims of water-supply. This inevitable abstraction of water from many rivers, however, should be compensated for, in the case of a tidal navigation, by a corresponding maintenance of the channel near the tidal limit by dredging, as the depth of a tidal river at its upper extremity depends upon the fresh-water discharge.

Nature and theory both indicate that the best form for a tidal channel or estuary is that of a trumpet, gradually enlarging in width as it approaches the sea, and having a more rapid increase in width near its outlet. This form ensures the freest admission of the flood tide, and the most complete filling of the tidal receptacle; and by promoting uniformity of flow, it secures uniformity in depth. A narrow outlet to a wide estuary produces a powerful scour, and consequently a considerable depth in the contracted channel; but the scouring influence of the flood tide above the outlet, and of the ebb tide below, only extends for a limited distance, depending upon the rise of tide and the extent of the estuary; and nature tends gradually to shape the channel to an expanding form by the deposit of alluvium in the wide sheltered recesses of the estuary, where the current loses its velocity.

CHAPTER XI.

DREDGING IN TIDAL RIVERS.

MANY tidal rivers, in their natural condition, are impeded by hard shoals of clay, gravel, boulders, or rock, which the scour of the current is unable to remove, even when the river is regulated by training works. Moreover, in the case of softer materials, a current is capable of maintaining a depth in a channel, which its scour is unable to obtain; whilst in some rivers leading to important centres of trade, a depth is required for the accommodation of sea-going vessels of large draught, which the current of the river would be wholly incapable of providing, or even of maintaining when artificially procured. The removal of hard shoals often effects a considerable improvement in a tidal river, for it facilitates the influx of the flood tide, and promotes the scour of the ebb, thus both increasing the tidal volume, and augmenting the depth over the softer shoals by the improved scour resulting from the greater velocity of the current. The dredging of soft shoals, moreover, not merely increases the depth of the channel, but also increases the tidal capacity of a river when the shoals are above low-water level, and even when they are below this level, in facilitating the flow by the enlargement of the channel, and by lowering the low-water line. The deepening of a channel, however, beyond certain limits, reduces so much the rate of flow, that the current deposits some of its sediment; and the depth has then to be maintained by dredging.

The improvement of most tidal rivers involves regulation and training works, as well as dredging; but in some cases the increase in depth has been principally effected by dredging, of which the Tyne, the Tees, and the Clyde are instances; whilst in other rivers, training works have been supplemented by large dredging operations, examples of which are furnished by the Ribble and the Weser. The reduction in cost of dredging, effected within recent years, by great improvements in dredging appliances, has led to a large extension of the system; and the certainty of its effects, and the facility with which the depth of a river can be increased by dredging in proportion to the development of a port to which the river gives access, have rendered dredgers almost indispensable machines for the improvement of rivers.

The success which has attended the efforts to improve the approach channels to the ports of Dunkirk, Calais, and Ostend, across the sandy foreshore encumbering their entrances, and also the deepening of Gedney's channel leading to New York harbour, by means of sand-pump dredgers, have resulted in the endeavour to lower the sandy bar of the Mersey, in Liverpool Bay, by a similar method. Accordingly, the employment of dredging for the improvement of tidal rivers, which has rendered such services in deepening their comparatively sheltered outlet channels, is now being extended beyond their mouths into the open sea.

RIVER CLYDE.

The Clyde, draining an area of 945 square miles, was formerly a small stream encumbered by shoals; but it flows into a deep, wide, and long estuary, called the Firth of Clyde, which extends from Greenock to the sea. Its outlet is therefore most effectually protected, so that no difficulty is experienced in maintaining it. Considering that the improvement of the outlet of a river, across a shallow beach and exposed to waves and littoral currents, is generally the most difficult problem in river improvement, the Clyde is peculiarly favoured in this respect (Plate 8, Fig. 7).

Training Jetties, and Removal of Hard Shoals. Till 1773, the Clyde was such an insignificant river that it was fordable on foot at Dumbuck Ford, more than twelve miles below Glasgow; and there were numerous shoals higher up[1]. In that year the system of narrowing the river by jetties was commenced, so that it might scour itself out a deeper channel; and hard gravel shoals, which the current was unable to act upon, were removed by dredging or ploughing. In the two following years, the channel across Dumbuck Ford was made 7 feet deep and 300 feet wide, at low water, by dredging; and in 1781, the natural scour of the river had lowered it to 14 feet.

About twenty years later, more than two hundred additional jetties were erected; and low rubble training walls were constructed connecting the ends of the jetties, to render the channel uniform in width, and to prevent the slipping in of the accumulated deposit on the banks, and the consequent formation of shoals. Also, as the original jetties had been constructed to no uniform lines, they were regulated by shortening several, and lengthening others, so that the river might enlarge uniformly as it approached the sea.

[1] 'The River Clyde,' James Deas, Minutes of Proceedings Institution C. E., 1873, vol. xxxvi, p. 124; and Transactions of the Institution of Naval Architects, July, 1888.

The widths then fixed for the river were, 180 feet just below the harbour of Glasgow, enlarging gradually to 696 feet at Dumbarton Castle. The increase, however, in the size and number of the vessels frequenting the port became subsequently so great, that the river has been widened to 370 feet at the lower end of Glasgow harbour, enlarging by degrees to 1,000 feet at Dumbarton Castle. The low training walls were gradually raised, as the ground behind them became higher, either by natural silting-up, or by the deposit of material dredged from the river; and the land was reclaimed. Quays were gradually extended along the banks of the river; and, from 1807, the channel was continuously deepened, widened, and straightened, for about thirty years.

Deepening the River Clyde by Dredging. In 1840, a scheme of improvement was authorized, in accordance with which the subsequent widening and deepening of the river have been carried out (Plate 8, Figs. 7 and 8)[1]. The present navigable condition of the river is, indeed, the result of the systematic dredging operations inaugurated at that period, and since then carried on without intermission with gradually improved and more powerful plant. In 1844–5, five bucket-ladder dredgers were at work in the river, capable of dredging to a depth of from 13 to 20 feet; and 234,000 cubic yards of material were removed in the year. The work proceeded at about the same rate during the next six years, till in 1852 a dredger working to a depth of 22½ feet took the place of the smallest of the original dredgers; and the amount raised during the five following years averaged about 350,000 cubic yards annually. A dredger capable of working to a depth of 28½ feet was added in 1856, so that in 1856–7 the quantity dredged reached 506,000 cubic yards, and the next year 630,000 cubic yards. The dredged material, up to 1862, was discharged into punts of 8 cubic yards capacity, and de-

---

[1] A recent plan and longitudinal section were furnished me by Mr. James Deas, the engineer to the Clyde Navigation Trustees.

posited on the foreshores and low-lying land adjoining the river; but since then the material has been conveyed in steam hopper barges, holding from 240 to 320 cubic yards, and deposited in deep water in Loch Long, an inlet from the Firth of Clyde. This method of disposing of the dredged materials has enabled much larger quantities to be extracted from the river; and by the gradual substitution of more powerful dredgers for the earlier machines, capable of dredging to depths of 28½ to 34 feet, necessitated by the increase in depth of the river, from one to over one and a half million cubic yards have been removed annually since 1874–5, a maximum of 1,841,300 cubic yards having been dredged in the year 1892–3, which was accomplished by five bucket-ladder dredgers, and a 10-ton steam digger. The total quantity of material dredged from the river and harbour in the last forty-nine years amounts to 41,352,800 cubic yards. The continuance of the deposit of the dredgings in Loch Long having been objected to as a nuisance, steam hopper barges of 1200 tons capacity, and having a speed of 10½ knots an hour, have been constructed for conveying the dredged materials further down the Firth of Clyde, to the new depositing ground three miles below Garroch Head.

A reef of hard whinstone rock was discovered in 1854, cropping up in the bed of the river 925 feet long and 320 feet broad; and it had to be bored and blasted, which was effected from a barge by diamond drills and dynamite, involving a cost of £70,000 for the removal of 110,000 tons of whinstone and boulder clay, so as to provide a depth of 20 feet at low-water spring tides over the reef.

Results of Dredging in River Clyde. By means of dredging, aided in the earlier stages of the improvement works by training and regulation, the Clyde has been converted from an insignificant stream into a deep navigable river, capable of giving access to ocean-going vessels of large draught up to Glasgow, where extensive basins and large graving docks have been constructed for their accommodation. A comparison of the longitudinal sections of the bed of the river in 1758 and 1891 (Plate 8, Fig. 8), indicates the great increase in depth that has been effected in the Clyde between Glasgow and Port Glasgow, along a distance of about 19 miles, amounting on the average to about 22 feet between Glasgow and Dumbarton. In 1758, the Clyde was only 15 inches deep at Glasgow at low water, 18 inches about 5 miles below Glasgow bridge, and 2 feet at Erskine and Dumbuck; whilst the rise of tide at Glasgow was only 2½ feet. At the present time, the bed of the river is practically level from Glasgow to Port Glasgow, affording a depth of from 17 to 20 feet at low water along the whole distance; and the rise of spring tides at Glasgow is 11⅛ feet, an increase of 8¾ feet, and 2 feet more than at Greenock. The high-water line has been raised about 3 inches at Glasgow since 1758, owing to the improved form and depth of the channel more than compensating for the increase in the volume of tidal water required for filling the river at each tide; and the actual rise of the high-water line between Greenock and Glasgow is nearly 2 feet. The great increase in tidal range at Glasgow is, however, due to the fall of the low-water line, resulting from the deepening of the river, which amounts to 8½ feet as compared with low water in 1758. This has naturally been accompanied by an advance in the time of high water at Glasgow, and an acceleration of the ebb tide. Thus whereas, in 1800, high water at Glasgow was two hours later than at Port Glasgow, it is now only one hour later; and whilst, in 1858, a float put into the river at Glasgow bridge just after high water, only reached Port Glasgow, a distance of 19½ miles, after being carried up and down by the tidal currents for 43½ tides, in 1881 floats under similar conditions travelled down 23½ miles in 10 tides on the average[1]. The

---

[1] 'Report on Tidal Velocities in River Clyde between Glasgow and Greenock,' James Deas, 1881.

removal, in 1880, of a weir which stretched across the river
a short distance above Glasgow bridge, has extended the
tidal influence higher up the river.

Till 1818, no sea-going vessels came up the river beyond
Greenock or Port Glasgow, their cargoes being discharged
into lighters for conveyance to Glasgow; in 1821 vessels
drawing 13½ feet came up the river; in 1850 vessels of 19 feet
draught could get up; and now steamers with a maximum
draught of 25½ feet can go up to Glasgow. The improvement of the Clyde has placed Glasgow in a foremost position
amongst British ports, as the tonnage of its vessels is only
exceeded by London, Liverpool, the Tyne ports, and Cardiff;
and the importance of Glasgow as a seaport has been largely
instrumental in making it the second city, as regards popu-
lation, in the United Kingdom. The deepening of the river
has also led to the development of shipbuilding on a very
large scale along its banks; and the recently built, gigantic
Atlantic liners, the Campania and the Lucania, 620 feet
long, 65 feet beam, and 26 feet draught, were launched on
the Clyde.

Remarks on River Clyde Works. The works on the Clyde
have extended over a long period; they have been carried
out under very favourable conditions in respect of the outlet;
and the deepening of the channel up to Glasgow has been
gradually increased, to keep pace with the growing requirements of the port, and the augmented draught of vessels.
The hard shoals which were lowered by dredging, and the
rocky reef which was removed by blasting, were natural
obstructions which prevented a deepening of the channel by
scour, and impeded the tidal flow; and their removal constitutes a permanent improvement in the river, facilitating its
maintenance as well as increasing its depth. The deepening,
however, of the bed of the river right up to Glasgow, to
the extent to which it has been carried, gives an artificial
character to the river, with a depth considerably in excess

of the power of the currents to maintain, and consequently necessitates its maintenance by artificial means. As the mouth of the river is situated on a deep and well-sheltered estuary, it is not exposed to the importation of sand stirred up by the waves from the bottom during storms, or drifting along the coast. Accordingly, the only deposit that can accumulate in the river comes from the alluvium brought down by the river, from the sewage and other refuse of Glasgow, and from the sandbanks opposite Port Glasgow and Greenock, which afford but a limited supply when compared with that of an open sea-coast. The depth, however, of the channel has been long ago carried beyond its natural limits, and it is constantly being increased to meet the requirements of navigation; so that each year the channel is becoming more artificial in its character, and its maintenance more costly. Consequently, a considerable portion of the materials which are being dredged from the river every year consists of deposit, which, if not periodically removed, would soon reduce the available depth of the river. The amount of deposit dredged from the river, which constitutes the work of maintenance, varies from year to year. In 1883–4, out of 1,040,700 cubic yards dredged, 703,000 cubic yards consisted of deposit; whilst in 1891–2, as much as 824,100 cubic yards of deposit were removed, and only 491,800 cubic yards in 1892–3. Probably from 600,000 to 700,000 cubic yards would have to be dredged annually from the Clyde merely to maintain its depth; and the dredging required for maintenance will increase with every augmentation of depth, unless the volume of sediment is reduced by diverting the sewage of Glasgow from the Clyde, an improvement which is urgently needed on sanitary grounds as well as for the sake of the river.

One incidental result of the straightening and deepening of the Clyde, has been the cessation of inundations from river floods in the low-lying parts of Glasgow; for the last time

that the river rose above the level of the quays of Glasgow
harbour, was in 1856. This immunity from inundations is
due to the facility with which the flood discharge passes
down the greatly enlarged bed of the river.

RIVER TYNE.

The river Tyne has been frequently compared to the Clyde,
because both the system of jetties and training walls, for
contracting and regulating the channel and concentrating its
scour, and also dredging, have been resorted to for its improvement. It differs, however, from the Clyde in some important
particulars. It was originally in a better condition than the
Clyde; and it is a somewhat larger river, having a basin of
1,130 square miles. Its course was not impeded by hard
shoals; but, instead of opening into a deep sheltered estuary,
it flows direct into the North Sea (Plate 8, Fig. 1). Newcastle, the port for which the river had to be improved, is
10½ miles from the sea-coast. The river between Newcastle
and the sea was very irregular in width, its navigable channel
was winding, shoals rose in some parts above low water in
the centre of the river, and a bar existed at its mouth.
Up to 1842 no improvements had been effected; and at
that period there was a depth, at the shallowest place in the
main channel, of only 2 feet at low water, and 14½ feet at
high-water spring tides; and the high-water level at Newcastle
was sometimes from 7 to 10 inches lower than at the sea.

Jetties and Training Walls. Between 1843 and 1858 the
river was systematically regulated by jetties, whose extremities were eventually connected by low training walls of
rubble and slag, after the spaces between them had been
partially filled up by the deposit of silt in the slack water[1].
These works were aided by a small amount of dredging
annually, commenced in 1838 with one dredger on a very

---

[1] Minutes of Proceedings Institution C. E., vol. xxvi, p. 406.

small scale, the amount raised being gradually increased, first by a more powerful dredger, and then by the employment of a second dredger in 1855, and a third dredger in 1857, whereby the yearly dredging was augmented from a minimum of 9,700 cubic yards in 1841, to 449,000 cubic yards in 1860. The minimum depth in the channel up to Newcastle, at high-water spring tides, was thereby increased to 18 feet, a gain of 3½ feet; high-water mark at Newcastle attained the same level as at sea; and the low-water line was lowered about 16 inches on the average. The amount of improvement effected by these works, up to 1860, was small, and had been accomplished slowly, but, being produced by natural scour, the river was sure to maintain its depth; and the cost of the works was moderate. The depth of the river, however, was far from satisfactory, as indicated by the longitudinal section of the river in 1860 (Plate 8, Fig. 2)[1]. The depth over the bar at the mouth was only 6 feet at low-water spring tides, and 21 feet at high water; the river also above was still greatly impeded by shoals; and the navigable channel was circuitous.

Tynemouth Piers. The converging breakwaters protecting the entrance to the Tyne, and facilitating the removal of the bar by dredging, have been referred to in Chapter IX (p. 213). They were commenced in 1856, with the object of deepening the approach to the river across the bar, being necessitated by the exposed state of the bar projecting beyond the coast-line in front of the mouth of the river. The breakwaters consist of a rubble mound, surmounted by a superstructure, with masonry facework and concrete hearting, and founded from 7 to 21 feet below low water[2] (Plate 8, Figs. 1 and 3). The entrance of 1,300 feet between the pierheads allows the flood tide to pass freely up the river. These piers,

---

[1] 'River Tyne Improvement,' P. J. Messent, 1888.

[2] 'Harbours and Docks,' L. F. Vernon-Harcourt, pp. 317–321, and plate 7, fig. 9.

though constituting an essential portion of the improvement works, have involved a greater cost than actually required for the navigation of the river, owing to their original limits having been extended for forming a refuge harbour.

Dredging in the River Tyne. The comprehensive scheme for the improvement of the river authorized in 1861, consisted mainly in deepening the river by very large dredging operations from its mouth right up to Hedwin Streams, about 8¾ miles above Newcastle, a total distance of 19¼ miles. In 1861 two new very powerful dredgers were set to work, in conjunction with two of the former dredgers; and two more were started in 1863, since which period there have been always five, and in most years six dredgers at work in the river [1] At the commencement of these operations, the amount dredged in a year rapidly rose from 1,243,000 cubic yards in 1861, up to a maximum of 3,515,700 cubic yards in 1866; since which time the volume of material raised from the river has been gradually diminished with occasional fluctuations, the smallest amount since 1861 having been raised in 1893, when 780,100 cubic yards were dredged from the river. The total volume of materials dredged from the Tyne and two docks, between 1838 and 1893, amounted to 62,309,000 cubic yards, out of which only 1,866,000 cubic yards were dredged in the twenty-three years preceding 1861; whilst in the last thirty-three years, 60,443,000 cubic yards have been removed, giving a yearly average of 1,831,600 cubic yards, almost equal to the total output during the earlier period. The dredging has been effected by bucket-ladder dredgers, loading into hopper barges which convey the dredged material out to sea between two and three miles beyond the mouth of the river, and deposit it in a depth of water of 20 fathoms at low tide. Three of the dredgers have been at work every year since they began work in 1843, 1855, and 1861 respectively, having each been only

[1] A statement of the annual quantities of dredging was furnished me by Mr. Messent, the engineer to the River Tyne Commission.

laid up once for repairs during periods of from four months to a year; and the maximum amount raised by one dredger in a year was 974,400 cubic yards, accomplished by the newest dredger in 1866, which, though not working in 1893, has raised the greatest amount of material, reaching a total of 14,553,000 cubic yards in its thirty years of work, including laying up on three occasions for repairs.

Dredging is being carried on in the Tyne at the present time, in order to maintain the depth already attained from the sea up to Newcastle and two or three miles above, and also for extending the increase in depth higher up the river to the limit proposed (Plate 8, Fig. 1).

Auxiliary Improvement Works in the Tyne. An abrupt contraction in the channel rather less than a mile above the bar, known as the 'Narrows,' which was an obstruction to navigation and impeded the flow of the tide, has been widened from 400 feet to 700 feet. The channel, moreover, has been rendered more uniform in width; and sharp bends inconvenient for navigation, have been eased by the removal of several projecting points, of which Bill Point and Whitehill Point were the most obstructive (Plate 8, Fig. 1). By the removal of Whitehill Point, the river has been widened 300 feet; and the rocky projection of Bill Point has been cut back 400 feet, involving the blasting of 380,000 tons of rock under water by means of dynamite, and the removal altogether of two million tons of rock and soil. This latter obstacle not merely produced a very sharp bend in the river difficult to steer round, but also, by its height of 72 feet, dangerously obstructed the view of any approaching vessel. The old stone bridge at Newcastle has been replaced by a large swing-bridge affording an enlarged waterway, thereby greatly facilitating navigation and the flow of the tide. Above Newcastle, the river has been regulated by training works, the channel having been widened in some places and narrowed in others, and points removed; and a straight cut has been substituted for a very circuitous channel near Blaydon, reducing the length by three quarters of a mile, and promoting the tidal influx.

Results of Works in River Tyne. The great improvement in the navigable condition of the Tyne between Newcastle and the sea, exhibited by a comparison of the longitudinal sections of the river in 1860 and 1888 (Plate 8, Fig. 2), amounting to an increase in depth of about 20 feet, has been almost entirely accomplished by the unprecedented dredging operations carried out in the river during the last thirty-three years. In 1860, in spite of a rise of nearly 15 feet at springs, vessels drawing more than 20 feet could not enter the Tyne at high-water spring tides; and vessels of 17 to 18 feet draught were detained sometimes for two or three months in the river during easterly winds, owing to the reduction in the available depth over the bar by wave oscillations. Moreover, only vessels with a draught not exceeding 15 feet could reach Newcastle, even at high-water spring tides, along a tortuous channel between sandbanks; whilst above Newcastle, small craft alone could navigate the river, and only during high water. At the present time, the depth at the outlet is about 24 feet at low water of springs; there is a deep piece of channel, 1½ miles long, near the mouth of the river, in which vessels can moor in 30 feet of water at low-water spring tides, waiting for an opportunity of going out to sea; and between Shields and Newcastle, where formerly steamers of only 3 to 4 feet draught used to ground for hours, there is now a depth of 20 feet throughout at the lowest tides[1]. The river has also been deepened for some three or four miles above Newcastle, to 18 feet below low-water spring tides; and this deepening will eventually be extended, at a reduced depth of 12 feet below the same level above Scotswood, to the proposed limit of the improvements at Hedwin Streams, about 4½ miles above Scotswood, and actually three-quarters of a mile beyond the limit of the tidal influence at spring tides in 1860.

[1] 'River Tyne Improvement,' P. J. Messent, 1888, p. 4.

The improvement of the Tyne has raised high-water mark at Newcastle about a foot, which is about the present rise of the high-water line between the sea and Newcastle, the line having been level in 1860; whereas it had a fall inland in 1842. The low-water line has been lowered about 3 feet at Newcastle since 1860; and this lowering had already reached a maximum of about 7 feet at Scotswood in 1885, which will be still further increased as the deepening progresses in the upper part of the river. It has been estimated that when the proposed deepening has been completed, the tidal capacity of the river will have been increased by 20 million cubic yards, which will very materially assist in maintaining the channel. The tidal range at Newcastle is now about 9 inches greater than at the mouth of the river.

High water at Newcastle, in 1860, occurred 1 hour 3 minutes later than high water at Shields; in 1872, the difference in time had decreased to 23 minutes; and now high water occurs at Newcastle only 12 minutes after high water at Shields. The tidal interval also between Newcastle and Newburn, which is about 7 miles higher up the river, has been reduced from 29 minutes to 8 minutes since 1860.

The deepening of the river has so greatly facilitated the passage of land floods down the river that the height attained by the floods has been reduced, so that the lands and quays bordering the river between Wylam and Newcastle are no longer subject to periodical inundation.

The great increase in the depth of the river has naturally exercised a powerful influence on the trade of the Tyne ports, which is manifested by the rapid increase in the yearly tonnage of the vessels entering and leaving the river, so that the Tyne ports come next in this respect to London and Liverpool amongst British ports. The average tonnage of the vessels has also gradually increased from 149 tons in 1854, up to 445 tons in 1886; whilst many vessels of between 2,000 and 4,000 tons enter the river now, which could not possibly have gained access to it in 1860. Shipbuilding also has been largely developed in consequence of the improved condition of the river.

Remarks on River Tyne Works. As in the case of the Clyde, the deepening of the Tyne, which has been mainly effected by dredging, has been carried considerably beyond the limits which the scour of the currents could maintain, notwithstanding the increase in the tidal capacity of the river. The depth, indeed, in which the outlet between the pierheads is situated, will prevent the introduction of much sand from the sea, except during violent storms ; and whatever sand may be brought by waves into the harbour will deposit in its sheltered area, and be readily removed. The alluvium, however, brought down by the river from inland, and the refuse thrown in from the banks, will deposit to a large extent in the deepened channel ; and therefore dredging will always be required for maintaining the depth attained, the average annual amount of which can only be determined after the completion of the deepening of the upper part of the river.

The river Tyne possesses the advantage of a well-defined channel, restricting the necessary works of improvement within narrow limits, and free from the uncertainties attending the improvement of the outlet channel of a river through a wide sandy estuary in an exposed situation. It furnishes an instance of a river in which dredging operations have been carried out to the largest extent within a distance of about 14 miles ; and though the deepening effected up to Newcastle is not greater than that on the Clyde near Glasgow, the amount of material removed is half as much again as in the Clyde, and has been mainly accomplished in two-thirds of the time. The large volumes of material, however, raised from the bed of the Tyne, vast as they appear, sink into insignificance when compared with the operations of nature ; for the Mississippi carries into the Gulf of Mexico in flood-time more material in ten hours than the maximum amount dredged out of the Tyne in a year.

RIVER TEES.

The river Tees, which has a drainage area of 735 square miles, was formerly very irregular and tortuous between Stockton and Middlesbrough. A short distance below the latter town it enters a wide sandy estuary, through which it flowed in a winding shifting channel for about 7 miles before reaching the North Sea (Plate 8, Fig. 4).

Straight Cuts and Jetties. The first improvement in the tidal portion of the river was made in 1810, when a cut, 220 yards long, was opened near Stockton, which reduced the length of the river by about $2\frac{1}{2}$ miles, and increased the depth at Stockton from 9 feet to 11 feet [1]. Another cut was made in 1830, having a length of 1,100 yards; but though removing a very bad bend, it did not shorten the river very much, whilst it cost twice as much as the first cut, and reduced the tidal capacity. About the same period the system of regulating and contracting the river by timber jetties, projecting out at right angles to the banks, was introduced, and was carried on till about 1852. By these means a deeper and more direct channel was formed between Stockton and Cargo Fleet below Middlesbrough, a distance of about 7 miles; the increase in depth varying from about 2 to 5 feet. The jetties, however, produced irregularities in the depth, shoaling the channel in the intervals between them, and merely deepening it in front of them. The works, moreover, by cutting off a large area of tide-covered strand between the jetties, which was raised by deposit, had a somewhat injurious effect on the channel just below. There were also shoals in the channel through the estuary, especially where it divided into two branches on each side of the Middle Sand, and also over the bar at the mouth of the estuary (Plate 8, Fig. 5).

[1] Minutes of Proceedings Institution C. E., vol. xxiv, p. 64.

Training Walls in the Tees. Owing to the deficiency in depth of the channel below Stockton, and the growing importance of Middlesbrough as the centre of a rapidly developing iron trade, the improvement of the river between Stockton and the sea by training walls and dredging was decided upon in 1853. Longitudinal training walls, composed of mounds of slag from the iron-works, were gradually formed on each side of the channel between Stockton and Middlesbrough, connecting the ends of the jetties and placed a suitable width apart. The channel was at the same time straightened and widened where expedient; two projecting points were cut off; and the foreshores at the back of the training walls have been reclaimed. The training walls, raised from 4 to 7 feet above low water, were by degrees extended through the estuary, as funds permitted, along the more direct southern channel; and the northern channel was shut off, thereby directing the currents into a single fixed channel[1]. These training walls regulated the bed of the channel, and rendered the channel stable, without materially increasing its depth except in soft places.

Deepening the River Tees by Dredging. Clay, with stones and boulders, underlies sand and silt in the greater part of the bed of the river, being met with at from 7 to 9 feet below low water; and a ledge of lias rock extended across the channel at the eighth buoy, so that two-thirds of the channel was laid bare at low water, leaving only a narrow gorge on the west side, through which the fall of the low-water line was considerably increased. Dredging, accordingly, and blasting the reef were essential for obtaining an adequate improvement of the river, for scour could not have removed the boulder clay; and the rocky ledge, besides presenting a dangerous shoal, impeded the influx and efflux of the tide.

The dredging has been effected by bucket-ladder dredgers, loading into hopper barges towed to sea by steam-tugs, which

---

[1] 'River Tees Improvements,' John Fowler, Minutes of Proceedings Institution C. E., 1887, vol. xc, p. 346; and 'The River Tees,' W. Fallows.

deposit the dredged materials 3 miles beyond the bar [1]. Dredging operations were commenced in 1854 with one dredger; and a second dredger was set to work in 1866, a third in 1870, a fourth in 1878, and a fifth in 1884. The total amount dredged from the commencement up to the close of 1880, was 6,380,000 cubic yards; and in the next five years, nearly the same amount was removed from the river by means of four double-ladder dredgers and a Priestman grab, the total reaching 12,371,700 cubic yards at the close of 1885, giving an average of nearly 1,200,000 cubic yards of dredging per annum during the five years. The total volume of materials raised from the Tees up to the close of 1893, reached about 19,784,000 cubic yards, an average annual amount of dredging of 926,000 cubic yards having been carried out during the last eight years; and the maximum quantity dredged in one year was accomplished in 1885, when 1,431,100 cubic yards were raised from the river [2].

The rocky ledge was removed, over an area of 13 acres, by means of diamond rock-boring drills and blasting with dynamite, 124,000 cubic yards being thus shattered, and the débris raised, procuring a depth of 14 feet at low water, at a cost of £26,780. The depth over the rock has now been increased to 20 feet below low water, by using cast-steel claws in a bucket dredger, alternating with two buckets on the ladder.

North and South Gare Breakwaters. The two breakwaters converging across the wide exposed outlet of the Tees estuary, have been already referred to in a previous chapter (p. 212). Their main object is to protect the estuary and the training walls, for the deepening of the outlet channel over the bar has already been effected by the tidal scour in the deepened channel directed by the training walls. Being

[1] 'Dredgers and Dredging on the Tees,' John Fowler, Minutes of Proceedings Institution C. E., 1884, vol. lxxv, p. 239.
[2] Recent particulars relating to the dredging were supplied by Mr. G. J. Clarke, engineer to the Tees Conservancy Commission.

chiefly composed of refuse slag from the neighbouring iron-works, they have been constructed at an unusually small cost (Plate 8, Figs. 4 and 6). Reclamations have already been made in the shallower portions of the estuary; and further reclamations are contemplated alongside the training walls, at the upper end near Cargo Fleet. These reclamations are favoured by the protection afforded by the breakwaters ; but if they are gradually extended, they will proportionately reduce the value of the estuary, as a sort of sluicing basin enclosed by the breakwaters, for maintaining the depth at the outlet. Gradual accretion of the estuary outside the trained channel will, indeed, result from alluvium brought down by the river, or sand driven in by storms; but any acceleration of this process should be avoided, for it will entail considerable increase in dredging for maintaining the depth near the mouth and over the bar outside.

Results of Works in River Tees. The great increase in depth obtained in the Tees since 1851, from the sea up to Stockton, as exhibited on the longitudinal section of the river (Plate 8, Fig. 5)[1], is the result of dredging, aided by the fixity of channel afforded by the half-tide training walls through the estuary. There was a depth of only 8 feet over the bar outside the mouth of the river in 1851, at low-water spring tides. Though no dredging has been carried out on the bar beyond the mouth, the improved scour into and out of the deepened channel has lowered the bar to 19 feet below low water; and a similar depth has been obtained by dredging for about 3 miles above the mouth of the river, except at two or three places where the depth is reduced to about 15 feet. As, however, the rise of spring tides is 15 feet, there is an available depth of 30 feet at high-water spring tides for over 3½ miles from the bar, being a gain

---

[1] A recent plan of the Tees was sent me by Mr. J. H. Amos, secretary to the Tees Conservancy Commission, and also a longitudinal section prepared by Mr. G. J. Clarke, the engineer to the Commission.

of about 10 feet as compared with the shallowest place in
1851. Proceeding up the river, there is a minimum depth
in mid-channel at low water, of 10 feet near Cargo Fleet,
8 feet opposite Middlesbrough, and 5 feet near Stockton,
the dredging having effected an average increase in the depth
of the bed since 1851, of about 10 feet up to Newport, and
about 5 feet from Newport to Stockton. It is proposed to
continue deepening the upper part of the river by dredging,
so as to obtain an available depth of 18 feet at low water
up to Middlesbrough, and of 12 feet up to Stockton.

High-water mark at Stockton is about a foot higher than at
the sea ; and the rise of spring tides at Stockton is 15 feet, the
same as at the mouth of the river. Since 1851, the low-water
line has been lowered 4 feet at Stockton by the improvement
of the river below ; and as the flood tide must have been
considerably impeded in flowing up the river in its former
condition, it is probable that high water did not rise as
high at Stockton formerly as it does now, so that the actual
increase in the range of spring tides there probably amounts
to 5 feet or more. A further lowering of the low-water
line may be anticipated as the deepening of the channel
up to Stockton progresses. The tidal flow and ebb have,
accordingly, been considerably facilitated in the channel up
to Stockton by the deepening of the river ; and the tidal
capacity of the channel has been increased.

Since 1878, vessels of from 1,000 to 3,000 tons have been
able to leave Middlesbrough fully laden and proceed to
sea drawing 21 feet of water. Eventually, when the proposed
extension of the depth of the navigable channel by dredging
shall have been accomplished, vessels of moderate draught
will be able to navigate the river between Middlesbrough
and the sea at any state of the tide ; and vessels of large
draught will be able to get up to Stockton at high tide.

Remarks on River Tees Works. The deepening of the
Tees has been effected mainly by dredging, as in the case

of the Tyne and the Clyde ; but the dredging operations have hitherto been on a more moderate scale, and the increase in depth considerably less. The proposed deepening, how-ever, has not reached such an advanced stage as on the other two rivers ; and the distance over which it extends is only 12 miles, as compared with about 19 miles on the Tyne and the Clyde. The removal of the eighth buoy scarp con-stitutes an important permanent improvement in the river ; but in spite of the improved tidal condition and increased tidal capacity of the channel, the deepening is being carried beyond the limits of natural scour, and every increase in depth will involve additional dredging for maintenance.

The lowering of the bar beyond the mouth of the Tees having been effected by the scour resulting from the regulated and deepened river, the maintenance of the depth over it is dependent on the permanence of this scour. The break-waters at the mouth of the estuary, by narrowing the outlet, tend, in the first instance, to intensify the tidal scour over the bar ; but they also, by the shelter they afford to the wide estuary, and the changes in velocity they necessarily produce in the tidal currents through the restricted outlet and over the wide estuary, favour accretion in the wide sheltered expanse, leading to eventual reclamation, and diminution in the tidal capacity of the natural sluicing basin. The reclama-tion of 2,600 acres has, indeed, been already effected along the shallow borders of the estuary (Plate 8, Fig. 4) ; but these portions added but little to the tidal capacity, owing to their insignificant depth at high tide ; whereas the sale of the reclaimed lands provided funds for the extension of the improvements. Further reclamations, however, of 1,000 acres are now proposed, situated on each side of the channel below Cargo Fleet ; and it is probable that these will be followed later on by other reclamations, especially when the completion of the North Gare breakwater increases the shelter in the estuary. Accordingly, the continued maintenance of the bar by tidal scour cannot be regarded as assured; but it is fairly certain that the income derived from the sale of the reclaimed lands will more than provide for the maintenance of the depth over the bar by dredging. Otherwise it would be expedient to abandon further reclamations, and to maintain the tidal capacity of the estuary by periodically removing, by dredging in the sheltered area, a volume of material equivalent to the amount of accretion.

RIVER USK.

The river Usk, having a drainage area of only 634 square miles, and flowing into the Bristol Channel about 4 miles below the port of Newport to which it gives access, has been far more favourably endowed by nature than the rivers just described. Opening at right angles into a sheltered portion of the Bristol Channel a long distance from the open sea, its outlet needs no protection; and though the flood tide comes into the Usk densely charged with the yellowish mud which the waters of the Severn estuary hold in suspension, and which readily deposits in the creeks, foreshores, and sheltered places of the river, the main low-water channel is generally fairly free from deposit, being scoured by the rapid tidal currents aided by the fresh-water discharge. The range of tide, moreover, in the lower part of the river is exceptionally large, attaining 40 feet at a high spring tide at the mouth, and 34 feet at Newport bridge 5 miles above, which constitutes the upper limit of the seaport (Plate 8, Figs. 9 and 10). The trumpet shape of the high-water channel of the river raises the high-water line of spring tides one foot higher at Newport bridge than at the mouth, in spite of the tortuous course of the river. As the river rises in a mountainous and rainy district, its fresh-water discharge is large in proportion to the size of its basin. Notwithstanding however, these favourable conditions which have enabled a large seagoing trade to grow up at Newport, the state of the river is not free from defects, which demand attention in view of the constantly increasing draught of vessels and requirements of shipowners, and the keen competition of the neighbouring South Wales ports.

Defects in Approach Channel to Newport. Though there is an ample depth in the river up to Newport at high-water spring tides for vessels of the largest draught, the depth in mid-channel at high water of the lowest neaps is reduced by some shoals in the middle of the river below the requirements of navigation, so that occasionally vessels of large draught have been delayed for two or three days from insufficiency of depth in the channel. The highest shoal is a sandbank which has accumulated below the old dock entrance, driving the low-water channel against the left bank of the river, over which the depth at the highest part, in the centre of the river, at high water of the lowest neaps, is only 17 feet (Plate 8, Figs. 9 and 10). This is the most defective portion of the river within the limits of the port; but as it is above the part traversed by the vessels of largest draught, which enter the Alexandra docks, a hard shoal composed of stone, compact clay, and boulders, known as the 'gravel patch,' lying in the centre of the river below the entrances to the principal docks, is a more important obstacle, though the depth over it at the lowest high tides is 22 feet. The other defects in the river consist in the marked divergence of the flood and ebb channels in three or four places, occasioned by the bends, and resulting in sandbanks in the middle of the river; the tendency of the ebbing current to cut into the concave bends, increasing the irregularities in the width of the river; and the hard shoal formed at the mouth of the river Ebbw, by the gravel brought down by this tributary and the refuse carried down in flood-time from the works along its banks.

Improvements proposed for the Usk. It was proposed in 1883 that the 'gravel patch' shoal should be removed, as not merely liable occasionally to impede the passage of vessels of very large draught, but also as preventing the natural scour of the river producing its full effect both above and below the shoal ; for its removal consequently affords a prospect of improving the condition of the river beyond the mere deepening of the channel at the shoal [1]. It was also urged that the injuries to the Usk caused by the accumulation in its channel of hard refuse brought down by its tributaries, the Ebbw, and the Afon Llwyd which flows in above Newport, should be arrested at their source, by preventing the manufacturers using these rivers as receptacles for their refuse. Stringent provisions for dealing effectually with this nuisance were, accordingly, obtained by the Newport Harbour Commissioners in their Act of 1890. Another recommendation made in 1883 was also carried out, namely, a series of cross-sections of the river between Newport bridge and the mouth, to enable future changes in the river to be readily ascertained.

By aid of the cross-sections taken in 1884, and some new ones over the same lines, the state of the river was thoroughly investigated in 1890 ; and as an improved depth over the shoals appeared more urgent than in 1883, a scheme was proposed for deepening the river by dredging [2], so as to provide a minimum depth in the main channel, at high water of the lowest neaps, of 24 feet from the bridge down to near the upper entrance to the Alexandra docks, and 26 feet for the remainder of the channel down to the mouth of the river (Plate 8, Fig. 10). This improvement of the depth was approved by the Commissioners ; and the proposed dredging is being carried out by a dredger with hopper barges built specially for the purpose. Training works also for regulating some wide reaches of the river, thereby securing

---

[1] 'Report on the River Usk and its Tributaries,' L. F. Vernon-Harcourt, 1883.

[2] 'Report on the Improvement of the River Usk,' L. F. Vernon-Harcourt, 1890 ; and 'The River Usk and Harbour of Newport,' L. F. Vernon-Harcourt, Brit. Assoc. Cardiff Meeting, 1891.

a better maintenance of the deepened channel, and preventing a further increase of the sinuous course of the river and of the variations in width, constituted a portion of the proposed works. The dredging required for obtaining the proposed depth is only about half a million cubic yards, which is a very small amount compared with the quantities dredged from the Tyne, the Clyde, and the Tees; but it is scattered, of small thickness, and some of it very stiff material; and it is probable that a greater depth will be considered expedient after the proposed depth has been reached. The lowering of the shoals will facilitate the influx of the tide; and by also promoting the more thorough efflux of the ebb, the low-water line, which at present has an irregular fall, will be lowered; and the difference of 6 feet between the range of tide at the mouth and at the bridge, will be materially diminished.

The river Usk is so well endowed by nature, that its improvement was not so urgent as on less favoured rivers; but a river with such natural capabilities cannot be allowed to fall behind other rivers in accommodation for navigation. Though the great range of its tides renders it inexpedient to attempt to make the Usk accessible at all states of the tide, like the three rivers previously considered, the works in progress will provide a channel of ample depth for some time before and after high water, of even the lowest tides, for vessels of the largest draught up to the principal docks.

RIVER MERSEY.

The river Mersey, draining an area, in conjunction with the Weaver, of 1,722 square miles, flows through an extremely irregular estuary into the sea. The river, which is narrow and winding near Warrington, gradually expands below, but is contracted at the rocky gap of Runcorn, below which the estuary again widens out, increasing rapidly in width below Hale Head, opposite which the Weaver flows in;

and attaining a maximum width of about 3 miles in front
of Ellesmere Port, it contracts again, and flows in a comparatively narrow channel, with a minimum width of about
three-quarters of a mile, between Liverpool and Birkenhead
(Plate 7, Fig. 11). The river emerges abruptly at New
Brighton into Liverpool Bay, and flows through sandbanks
and over a bar into the Irish Sea. The depth of the low-
water channel exhibits similar variations, for it is shallow
and variable down to Garston, below which it deepens to
a minimum depth of about 50 feet in the narrow channel
past Liverpool, and gradually shoals through Liverpool Bay,
till in passing the crest of the bar, the depth has sometimes
been reduced to about 9 feet below the lowest low water
(Plate 7, Fig. 12). The flood tide enters the estuary across
the bar, and also to a smaller extent from the west along the
narrow Rock channel, which is prevented from becoming
deeper by a ledge of rock across its entrance to the main
channel. Above the narrow neck, the flood tide pursuing
a straight course, maintains a deep blind channel in the
direction of Eastham along the Cheshire shore. The ebb
channel, on the contrary, whilst constantly shifting its course
in the upper part of the inner estuary, always emerges near
Garston on the Lancashire shore, and following along that
shore, only joins the flood-tide channel between Dingle Point
and Pluckington Bank.

Influence of Inner Estuary on Mersey Outlet Channel.
The wide inner estuary, together with the neck, has an area
of 22,500 acres, of which 17,300 acres become bare at low-
water spring tides; and it forms a vast natural sluicing basin
which, when filled during an equinoctial spring tide rising
29 feet at the bar, has a tidal volume of 710 million cubic
yards flowing out past New Brighton on the ebb, and 281
million cubic yards at a low neap tide. This large volume
of tidal water, being directed and concentrated by the narrow
neck in flowing into Liverpool Bay, maintains the outlet

channel through the sandbanks, which gradually shoals as the outflowing current spreads out with a reduction in velocity; and the sands are only prevented from forming a continuous strand across the bay by the tidal ebb and flow, for the fresh-water discharge of the river only amounts on the average to about 2,500,000 cubic yards per tide, which could only maintain a comparatively insignificant channel. The tidal capacity of the inner estuary is to some extent preserved by the slight preponderance the fresh-water discharge gives to the ebb, but mainly by the changes in the channel which, by fretting the sandbanks periodically, prevent their consolidation and growth, and by stirring up the sand facilitate its removal by the current. The inexhaustible supply of sand in the bay, and the prevalence of north-westerly winds blowing into the estuary, together with the difference in the velocity of the current in the neck and above, render it probable that most of the sand which encumbers the inner estuary has come from outside; and as in this instance the maintenance of the outlet channel depends on the maintenance of the tidal capacity of the inner estuary, the possibility of a continuance of the accumulation of sand inside is not devoid of interest. If the process is still in operation, it is a slow one, for the decennial surveys of the estuary, though exhibiting fluctuations, do not indicate any marked change; but they are not exact enough, nor have they been carried on long enough, to decide the question. The depth over the bar has remained fairly uniform for several years, exhibiting an average of about 10 feet below low water of the lowest tides, with slight variations which may be accounted for by changes in the position and in the cross-section of the channel; and the area of the cross-section of the channel, rather than the maximum depth, would be the true measure of the scouring capacity of the ebb current.

The bar appears to have gradually shifted seawards between 1840 and 1880, indicating a strengthening of the

ebb current, which must have been due to the regulation of
the narrow channel by the formation of quays on each side,
and also to the deterioration of the Rock channel, thereby
checking the diversion of a portion of the ebb in that
direction. Any increase in the tidal capacity of the inner
estuary would have had a similar effect; but beyond the
temporary influence of a large fret, further accretion is more
probable than erosion. Since 1880, however, a retrograde
movement of the bar seems to have commenced, for the bar
in 1890 was more than half a mile inside of its position in
1880, and approaching the place which it occupied in 1870
(Plate 7, Fig. 12).

Objections to Training Works in Inner Mersey Estuary.
Above Garston, the Mersey is quite unsuitable for vessels of
large draught, on account of the wanderings of the channel
and the inadequacy of its depth. The channel might be
fixed by training walls, which would increase the depth across
the sands by concentrating the scour; and it could be
deepened to any extent by dredging. This was the scheme
proposed for providing an outlet for the Manchester Ship-
Canal below Runcorn, in 1883 and 1884. This is the system
of improvement that has been adopted with success in the
estuaries of the Clyde and the Tees; and it might have been
adopted in the Mersey estuary if Liverpool and the bar
outside had not existed. There was no bar, and little
prospect of accretion at the mouth of the Clyde. On the
Tees, the training works have been carried nearly out to
the bar, and the outlet has been protected by breakwaters;
but in this case, if sand comes into the estuary, in spite of
the projection given to the outlet, accretion will take place,
and the bar will be raised. The training works, however,
in the inner Mersey estuary would have left the sands free
to be carried in from outside as before by the rapid flood
tide through the neck; and some of this sand, being brought
into the sheltered spaces on each side of the trained channel,

would have been deposited beyond the scouring influence of the ebb, which would have been mainly confined within the fixed channel. The sandbanks would, accordingly, have been gradually raised on each side of the estuary, without chance of disturbance by any wandering of the channel, thereby greatly reducing the tidal capacity of the inner estuary. The consequent reduction of the volume and velocity of the outflowing current would have inevitably led to the shoaling of the outlet channel, the raising and widening of the bar, and the shifting of its position somewhat landwards. Accordingly, independently of the paramount interests of Liverpool, the scheme was much less favourable for Manchester than the canal skirting the estuary down to Eastham, which has been carried out.

Dredging the Mersey Bar. The bar of the Mersey differs from the bars which existed in front of the Tyne and the Tees, in being much further out from the coast, being about 11 miles beyond the actual mouth of the river at New Brighton, owing partly to the large mass of sand encumbering Liverpool Bay, and also to the powerful ebbing current from the inner estuary. This distance, and the situation of the bar in relation to the coast-line preclude the possibility of resorting to converging breakwaters for lowering the bar. A training wall across the sandbanks, in continuation of the left bank of the river beyond the Rock lighthouse, would undoubtedly concentrate the flood and ebb tides in the main channel by closing the Rock channel and other subsidiary outlets, and consequently lower the bar; but such a work would be costly on shifting sandbanks and in an exposed situation. The bar, however, prevents Liverpool being accessible at all states of the tide; and Atlantic liners, which have crossed the ocean at a high speed, are liable to be delayed outside the bar if arriving near low water.

Since 1876, the approach channels to Dunkirk and Calais, across the sandy foreshore in the open sea, have been greatly improved by dredging with sand-pump dredgers; and more recently a similar system has been adopted with success at Ostend. The results achieved in dredging sand in the sea at these places and elsewhere, suggested the expediency of giving the system a trial on the Mersey bar ; and in September, 1890, a sand-pump dredger was set to work on the bar. Two sand pumps were at first fitted up on two steam hopper barges, capable of filling the hoppers of 500 tons capacity in an hour, and dredging to a depth of 36 feet[2]. These two dredgers succeeded in augmenting the depth over the crest of the bar, in the centre line of the navigable channel, from 12 feet to $18\frac{1}{4}$ feet below low water of the lowest tides, by the beginning of 1892, by the removal of 657,000 tons of sand, equivalent to about 438,000 cubic yards. During the next year and a half, however, though the dredgers removed nearly three times this quantity, the depth remained stationary, partly owing doubtless to sands from the sides being drawn down the steepened slopes, and partly to deposit in the deepened channel, which will always require periodical removal. Accordingly, a sand-pump hopper dredger was specially built for the work, capable of conveying 3,000 tons of sand in its eight hoppers, which can be filled in three-quarters of an hour[3]. The vessel can convey its load to the depositing ground at the rate of 10 knots per hour; and it commenced working in the latter part of 1893. The greatly increased dredging power thus obtained, aided by the greater stability of the large vessel, should enable the desired depth, of 26 feet over the bar at low water of the lowest tides, to be attained, and the deep channel enlarged to an adequate width, so as to give access to vessels of large draught at all states of the tide. Till the deep channel across the bar has been formed, and maintained for two or three

---

[1] Minutes of Proceedings Institution C. E., vol. lxxxix, p. 35.

[2] 'Dredging Operations on the Mersey Bar,' A. G. Lyster, London Maritime Congress, 1893, Sect. I, p. 50.

[3] 'The Sand Dredger "Brancker,"' A. Blechynden, London Maritime Congress, 1893, Sect. III, p. 52.

years, it will be impossible to tell how much sand will have to
be annually raised to preserve the depth, and consequently
what the cost of maintenance will be.   Besides, however, the
important deepening of the channel over the bar, which is sure
to be effected by the dredging plant at present in operation
in Liverpool Bay, the improved channel will facilitate the
influx of the flood tide up the navigable channel, and will
check the changes in the position of the main channel.

Concluding Remarks.   The foregoing examples of the
results of dredging operations demonstrate the great improvements that can be effected by dredging in the navigable
condition of tidal rivers, far exceeding the limits of depth
obtainable by natural scour directed by training walls.   Whilst,
however, natural scour is sure to maintain the depth which
it has formed, and may be promoted by the removal of hard
shoals and other obstacles which impede its action, and by
an extension of the tidal limit, dredging soon carries the
depth beyond the limits of natural maintenance, in spite of
the increase of the tidal volume produced by the lowering
of the low-water line, and therefore involves a continuance
of the dredging for maintaining the depth.   In seaports on
tidal rivers, which form centres of trade, the increase in traffic,
which generally results from an increased depth in the river,
and greater facility and safety of access, usually suffices to
pay for the expenditure on the deepening and the costs of
maintenance.   Moreover, in the present day, the continued
prosperity of a port depends upon its providing the increased
depth and accessibility demanded by shipowners; and many
ports are indebted to dredging for enabling them to maintain
their position and to extend their trade.

CHAPTER XII.

TRAINING WORKS IN ESTUARIES.

Simplicity of Regulation Works compared with Training Works in Estuaries. Influences of Training Works in Estuaries; Improvement of Trained Channel; Concentration of Scour; Accretion promoted outside Channel. *Training Works in the Wash:* Condition of Rivers draining Fens; River Witham, Training Works, New Outlet Channel, Effects of Works; River Welland, Training Works, Necessity for Prolongation; River Nene, Straight Cut, its Prolongation, Training Works, Need of Extension; River Ouse, Eau Brink Cut, Straight Cut below King's Lynn, Training Works in Estuary; Remarks on Improvement of Outfall of these Rivers. *River Dee Training Works:* Original Condition of Estuary; Training Works in Estuary, Reclamation the Main Object, Unfavourable Line adopted, Effects of Works; Remarks on Works, their Injurious Effect on Estuary. *Ribble Estuary Training Works:* Former Condition, Changes in Channels and Outlet; Training Works, their Extension, and Improvements in Channel; Works in Progress and Proposed; Effects of Works on Condition of Estuary: Remarks. *Weser Training Works:* Description of River and Estuary; Training Works in Tidal River and Estuary; Results; Remarks on Improvements and Prospects.

WHERE the channel of a tidal river is fairly well defined, as in the upper portion of most tidal rivers, and in rivers which flow direct into the sea, such as the Tyne and the Nervion, training works are necessarily limited to a regulation of the existing channel as regards its width, so as to secure a greater uniformity in depth, aided sometimes by a rectification of its course at sharp bends. The training works carried out on the Nervion below Bilbao, on the Maas below Rotterdam (Plate 6, Figs. 12 and 18), and on the Tees between Stockton and Middlesbrough (Plate 8, Fig. 4), regulated more or less defined, though irregular channels, giving them a gradually expanding width seawards, and a somewhat straightened course, so far as the existing conditions rendered reasonably practicable. A slight reduction in the tidal area of a river may be occasioned by such works, owing to the narrowing of the channel at unduly wide places; but this diminution in tidal capacity is generally compensated for by the progress upwards of the flood tide being facilitated. and by the lowering of the low-water line from the increased depth resulting from the training works. Moreover, an improvement in depth and an increase in tidal capacity can readily be obtained by dredging in the trained channel. When, however, a river emerges abruptly into a wide estuary encumbered with sandbanks, through which it flows in a winding, shallow, and shifting channel to the sea, the improvement of its outlet channel by training works is a far more complicated problem.

Influence of Training Works in an Estuary. Longitudinal training walls formed of mounds of rubble, slag, or fascines, guiding the channel of a river through a sandy estuary, fix the formerly wandering channel, and by concentrating the fresh-water discharge and the main flow of the flood and ebb tides into the trained channel, increase the scour and thereby improve the depth. Accordingly, training works fix, straighten, and deepen the channel of a river through a sandy estuary; and if their influence ended there, they would be unquestionably of great value in guiding rivers through sandy estuaries. In concentrating, however, the main currents within the trained channel, they withdraw the scouring influences of these currents from the rest of the estuary; and by stopping the wanderings of the channel, they prevent the periodical erosion of the sandbanks, which formerly prevented these banks from becoming consolidated and increasing in size. Any materials brought down by the river from inland, or stirred up by the waves and introduced by the flood tide, tend to deposit in the comparatively slack water at the sides of the estuary behind the training walls; and when once deposited, they are no longer liable to be disturbed by the shifting of the channel. Training works, consequently, promote accretion in the portions of an estuary outside the trained channel, wherever the waters coming into the estuary are charged with alluvium ; and this influence extends to the sides of an estuary below the ends of the training walls, where the works terminate within the estuary, owing to the reduction in the tidal capacity of the upper part of the estuary by the accretion produced behind the training walls. Experience, also, proves that, whilst the accreting influence of training walls may extend a considerable distance beyond their extremities, their influence in deepening the channel by the concentrated scour ceases a short distance below their termination. The rate of accretion in an estuary through which a river is trained is dependent upon the volume of alluvium introduced into the estuary, and the height of the training walls ; but even low training walls, in so far as they fix and deepen the channel and concentrate the scour, promote accretion at the sides of an estuary.

Wherever, therefore, alluvium is brought into an estuary, the benefits of training works to navigation, in fixing and deepening the outlet channel, are necessarily accompanied by a reduction in the tidal capacity of the estuary from the resulting accretion, which generally is only very partially compensated for by the increased tidal capacity of the deepened channel. In some cases, the diminished tidal capacity of the estuary is immaterial compared with the advantages of an improved channel ; but in other cases, the loss of tidal capacity imperils the maintenance of the depth at the outlet, or the accretion may prejudice the access to ports on the estuary. The local benefits of training works have often been exclusively kept in view in designing these works, whilst their effects on the estuary have been entirely overlooked ; so that training works have in some instances been attended by totally unexpected results. Accordingly, before commencing training works in an estuary, the probable accretion and its consequences should be weighed against the advantages of an improved channel.

The Clyde is so protected from the introduction of alluvium from the sea, and the materials brought down by the river are so effectually removed by dredging, that the training works in its estuary have been wholly beneficial; and any accretion that may have occurred behind the walls has been far more than compensated for by the improved tidal capacity of the dredged channel. The Tees estuary is not so secure from accretion, in spite of the protection afforded by the breakwaters at its outlet; but any loss of tidal scour that may gradually result from the training works, can be made up for by dredging. In some other estuaries, however, the balance of advantages is not so manifest; and under any circumstances, the consequences of the accretion resulting eventually from training works, as well as the immediate benefits to the channel, should be fully considered at the outset.

TRAINING WORKS IN THE WASH.

Four rivers draining the extensive Fen districts, flow into the Wash, which is a sort of wide estuary or inlet, encumbered with sandbanks, enclosed by the Lincolnshire and Norfolk coasts. These rivers, the Witham, the Welland, the Nene, and the Ouse, draining areas of 1,079,·760, 1,077, and 2,981 square miles respectively, traverse very flat low-lying districts for several miles before reaching the Wash, and then pass through tortuous channels between the sandbanks into deep water in Boston and Lynn deeps. The small fall of the land throughout the Fen country necessitates the supplementing of the straightened and embanked rivers by numerous straight drains, for the discharge of the inland waters; and the low level of the lands has led to the closing of the outlets of some of the main drains against the tide by gates at sluices to prevent the inroad of the sea, as for instance at the Black sluice, Maud Foster sluice, and Hobhole sluice on the Witham, the North Level sluice on the Nene, and the Middle Level sluice on the Ouse. Protection, moreover, was further provided for this district against the sea by the simple expedient of· arresting the flow of the tide at the Grand sluice just above Boston on the Witham, and at Denver sluice on the Ouse, by gates pointing seawards, without due consideration of the influence of this exclusion of the tide on the outfalls of these rivers. The restriction of the flow of the tide by these sluice-gates, together with the diversion to some extent 'of portions of the fresh-water discharge from parts of the main channels ·by the supplementary drains, produced a deterioration in the tidal portion of these rivers. Training works, however, have been carried out on these rivers for deepening their outlet channels, and improving the condition of the ports to which they give access. The average rise of spring tides in the Wash is 23½ feet.

River Witham. The outfall channel of the Witham between Boston and Hobhole, a distance of three miles, was trained and straightened nearly seventy years ago by banks of fascines, stones, and earth, in a uniformly increasing section seawards. This work improved the channel so much, that the flood tide reached Boston 1½ hours earlier; and vessels of larger draught could get up to the haven More recently the training works were continued towards the junction of the channel of the Witham with the Welland. The direction, however, of the Witham outfall till it joined the Welland, was at right angles to its subsequent course, which followed the line of the Welland outfall into Boston deeps. This circuitous channel, occasioned by the hardness of the strata of a projecting piece of foreshore known as the Scalp, was shallow and very inconvenient for navigation, and not only checked the progress of the flood tide, but also impeded the discharge of the land waters; and the improvement of the outfall, by opening a new channel for the river through the Scalp, was recommended at various times.

In 1880 the proposed outlet channel for the Witham across the Scalp was at length commenced; and it was completed in 1884. This curved channel, 3 miles long, 100 feet wide at the bottom at Hobhole, and increasing to 130 feet at its outlet at Clayhole at the head of Boston deeps, was excavated through silt, clay, peat, and boulder clay [1], reducing the length of the channel between Hobhole and Clayhole by nearly one half. The ends of the new channel were guided and secured by training walls of fascines with clay, weighted at the top by rubble stone. The river between the Grand sluice and the new channel was at the same time improved by training and dredging, affording an increase in depth of 8 feet, and a navigable depth of 22 feet at high-water spring tides up to Boston dock. The new channel has caused the flood tide to commence flowing up at Boston an hour earlier at spring tides than formerly; and as the tide no longer rushes up with a bore charging the river with sand in passing over the shallows, the deposit at the Grand sluice, which used to attain a height of over 11 feet at the close of a dry summer, is now insignificant. The facilitating also of the efflux of the ebb by the improved outfall has produced a lowering of the low-water line, amounting to 5½ feet at Hobhole and 4 feet at the Grand sluice, thereby materially increasing the tidal ebb and flow, for spring tides formerly rose only 12½ feet at the Grand sluice; but unfortunately for the outfall, the tidal flow of the Witham has been permanently restricted to the nine miles between deep water and the Grand sluice. The old outfall channel was only suitable for vessels of 300 tons; whereas the new channel gives access to vessels of 2,000 tons.

River Welland. The outfall of the Welland was much improved by training the river with fascine work for 3½ miles

[1] 'The Witham New Outfall Channel,' J. E. Williams, Minutes of Proceedings, Institution C. E., 1889, vol. xcv, p. 81.

below Fossdyke Bridge, which lowered the low-water level
7 feet at the bridge ; but want of funds prevented the continuation of this work down to its former junction with the
Witham.   The new cut for the Witham across the Scalp has
diverted the outfall of this river away from the Welland,
so that the two rivers only form an uncertain junction now at
Clayhole ; and the Welland has lost the scour of the Witham
along two or three miles of its outfall channel through the
estuary.   Accordingly, in order to improve the outfall of the
Welland, it would be necessary to extend the training works
in a fairly straight line to join the new outlet of the Witham
at Clayhole.   Besides deepening and fixing the Welland outfall and facilitating its tidal flow, the prolongation of the training walls would also be advantageous to the Witham in
directing the flow of the Welland into Clayhole, thereby
assisting in maintaining the outfall of both rivers into Boston
deeps ; whereas at present the Welland is liable to be diverted
into a more southern channel which at one time it followed.

River Nene.   In 1721, Kinderley proposed that a straight
cut should be made below Wisbeach, towards the sea, for
improving the outfall of the Nene.   This cut was only commenced in 1770, about five miles below Wisbeach, and was
carried out for a length of 1½ miles.   Though this cut was
considerably shorter than proposed, its effects sufficed to show
that the principle was correct, and that important benefits
might result from an extension of the system further down
the river.   The improvement effected by this cut was
gradually lost ; but in 1827–30, the cut was prolonged for
five miles down to Crabhole, being only partially excavated,
and was scoured out to the requisite size on letting in the
river.   The scour, in fact, of the fresh and tidal waters combined, was so effectual, that the sides of the channel had to be
protected by stonework.   The tidal flow along the river was
much improved by the cut, so that its effects extended as far
up as Wisbeach, five miles above the cut, to such a degree

that buildings had to be protected from being undermined ; and the low-water line was lowered more than 10 feet.

The Nene has been trained for a short distance into the estuary; but beyond the end of the training works, the channel. is shallow and tortuous ; and it would be necessary to extend the training walls for three or four miles to reach deep water in the Wisbeach channel leading to Lynn deeps.

River Ouse. This river was considerably improved above King's Lynn by the formation of the straight Eau Brink cut, in substitution for a very tortuous channel, which was completed in 1821. This cut lowered the low-water level about 6½ feet at its upper end; and it admitted a sufficient flow of tide to scour the river up to Denver sluice, enabling the sill of that sluice to be lowered 6 feet, which greatly benefited the drainage of the district.

The river has since been improved below King's Lynn by a straight cut two miles long, and by training the river for an additional distance of 1½ miles through the sandy shoals, with training walls of fascines, which works have effected a further reduction of 3 feet in the low-water level. The channel, however, beyond the training works is circuitous and somewhat shallow ; and an extension of the training walls for about three miles would materially improve the outfall channel.

Remarks on Outfalls of Fen Rivers. There are only two methods of improving the tidal portions of the Fen rivers flowing into the Wash, namely, straight cuts and training works ; for the removal of the barriers formed by the sluices is precluded by the interests of drainage. Straight cuts increase the fall of these rivers, which is very defective ; they consequently afford a freer discharge to the drainage waters which, accordingly, have a greater scouring force; and they facilitate the tidal flow up and down, which assists in the maintenance of the outfall.

Training works through the sandbanks in the estuary have also a beneficial effect both on navigation and inland drainage, for they straighten, fix, and deepen the outlet channel ; and by facilitating the influx of the tide, they promote the mainten-ance of the outlet, and by lowering the low-water line they facilitate the discharge of the inland waters. The extension, however, of the training works promotes accretion at the sides, and facilitates reclamation ; and as the estuary exhibits indica-tions of a gradual silting-up along its shores[1], and reclamations of the foreshores are made from time to time, attention will always have to be directed to the maintenance of the depth at the outlets of these rivers ; and the training walls will eventu-ally have to be prolonged as accretion progresses.

RIVER DEE TRAINING WORKS.

Original Condition of Estuary. The river Dee, draining an area of 813 square miles, flows into the Irish Sea through a wide open estuary which formerly extended up to Chester. As the estuary faces the north-west, it is exposed to the intro-duction of the sands which border the adjacent coasts during storms from that quarter ; and it is, consequently, encumbered by extensive sandbanks. Accretion appears to have been taking place in the upper part of the estuary, along the Flint-shire shore, more than two centuries ago, which was subse-quently assisted by the enclosure of the marsh lands. There was, however, a good wide and direct channel near the Cheshire shore in 1684 (Plate 9, Figs. 13 and 14), which afforded a depth of 6 feet at low-water spring tides opposite Burton Point, increasing to 12 feet opposite Heswell, and reaching 60 feet off Air Point at the mouth of the estuary, which with a rise of tide at springs of 29 feet at the outlet, and about half that height at Burton Point, provided a good depth at high water for vessels of that period up to this part of the estuary[2]. The channel, nevertheless, was naturally unstable in the sandy

[1] Minutes of Proceedings Institution C. E., vol. xxviii, p. 109.
[2] 'Admiralty Inquiry, Chester, Dee Navigation Improvement,' 1849, Drawings A and B.

estuary; and by 1732, the progress of accretion along the
Welsh shore, above the site of Connah's Quay, had driven the
channel close to the Cheshire shore above Burton Point, resulting in the formation of a tortuous channel round that rocky
point.   Whilst the channel below retained its ample depth and
width, the channel between Chester and Burton Point showed
signs of deterioration ; and the available depth up to Chester
at high-water spring tides was only 9 feet.   Accordingly, it
had become evident that works were needed in the upper part
of the estuary to improve the channel, in order for Chester to
retain its position as a seaport.

Training Works in the Dee Estuary.   The flood tide entering the estuary mainly from the west, formed a deep channel
over towards the Cheshire shore; and so long as the main
channel remained near the Cheshire shore, the ebb tide
followed the flood-tide channel, and a deep wide outlet was
maintained.   Any training works in the estuary should mani-
festly have been designed to preserve this favourable condition,
following approximately the channel of 1684, and thereby
avoiding the diversion of the channel at Burton Point produced
in 1732, and forming a straighter course for the river near
Chester.   At this period, however, a company was constituted,
called The River Dee Company, which, in consideration of
providing a channel 15 feet in depth at high-water spring tides
up to Chester, was authorized to reclaim the estuary down to
Burton Point, about 7,000 acres in extent.   The upper part
of the estuary was thus practically handed over to a land
reclamation company, without due consideration on the part of
the persons interested in the port of Chester, that the outlet
channel was liable to be compromised by the proposed works.
Though ostensibly proposing to improve the navigation up to
Chester, the works were carried out with a view to the most
profitable arrangement for the reclamation of the land, the real
object of the company.   Accordingly, instead of following the
course indicated by nature near the Cheshire shore, and

improving the tortuous course just below Chester, the river was trained between parallel banks like a canal ; the defective course near Chester was made worse, by training the river at a very acute angle to secure more land ; and then after another bend at Saltney directing the channel over to the Flintshire shore, the training works were terminated close against the shore at Connah's Quay (Plate 9, Fig. 13). This arrangement left the whole area of the upper estuary between the trained channel and the Cheshire shore available for reclamation, so as to form an undivided estate ; whilst the river provided a well-defined boundary between the Dee Company's land and the marshes on the Flintshire side. The training works appear to have been commenced about 1737, and to have been completed by 1771, at which period 2,400 acres of land had been reclaimed from the estuary by the company ; by 1826 the area reclaimed reached 4,000 acres ; and the remaining 3,000 acres are in process of reclamation. Altogether, including the reclamations by the landowners on the Flintshire side, the tidal area lost to the estuary on the completion of the last reclamation, will amount to over 10,000 acres.

The trained channel, besides being diverted from its natural course, directed the issuing current towards the projecting Pentre rock near Flint, which turned the channel across the estuary, as indicated in the charts of 1771 and 1849. The prolongation of the outer training wall for about $1\frac{1}{2}$ miles towards Flint, in a direction diverging from the shore, has made the channel pass outside the Pentre rock ; but it still tends to go across the estuary lower down, though its connexion with the flood channel on the Cheshire shore has now been cut off at low water[1] (Plate 9, Fig. 13). The training walls, formed of embankments protected by stone, improved the depth of the channel between Chester and Connah's Quay, but did not secure the stipulated depth of 15 feet at

---

[1] A chart of the estuary of the Dee in 1892 was furnished me by Mr. Enfield Taylor, engineer to the River Dee Commission.

high-water spring tides ; and occasionally at a low stage of
the river, the depth was reduced to such an extent that, even
at springs, vessels drawing only 8½ feet could not get up the
river[1]. The depth, however, of the trained channel has now
been increased, by a reconstruction of portions of the south-
west training wall, and by dredging, to a minimum depth of
15 feet. The channel, however, in the estuary beyond is
very tortuous and shifting; and in places between the end
of the training works and Flint, it is only 16¼ feet deep at
high-water spring tides (Plate 9, Fig. 14).

Remarks on the Dee Works. In the Dee estuary, exposed
to the inroad of sand from outside, and already showing signs
of silting-up in the seventeenth century, reclamations in the
estuary should have been sedulously avoided; and when
training works became necessary to maintain the depth of the
navigable channel to Chester, they should have been kept as
low as possible, and directed, with an expanding width seawards, in as straight a course as practicable near the Cheshire
shore. The flood and the ebb tides would thus have been
kept in the same channel, tending to preserve the good depth
below Burton Point; and any necessary increase in depth in
the trained channel could have been obtained by dredging.
The direction, and almost uniform width actually adopted for
the trained channel, were most unfavourable for navigation,
and for the maintenance of the channel by the aid of tidal
scour; whilst the wholesale system of reclamation undertaken,
by annihilating the tidal capacity of the upper estuary, has
occasioned a great growth of sandbanks in the lower estuary
from the loss of tidal scour, and a considerable reduction in
the depths of the channels (Plate 9, Figs. 13 and 14). In
fact, if it had been desired to destroy the navigable capabilities of the estuary of the Dee, it could hardly have been
accomplished in a more effectual manner than by the works
actually carried out by the River Dee Company. The history

---

[1] 'Second Report of the Tidal Harbours Commission,' 1846, p. 10.

of the Dee estuary serves as a warning of the fatal consequences that may result from mixing up schemes of land reclamation with the improvement of navigation in a sandy estuary.

RIBBLE ESTUARY TRAINING WORKS.

The Ribble estuary formerly commenced a very short distance below Preston; and about three miles further down, it commenced expanding rapidly to its existing wide outlet into the Irish Sea. The river at its mouth drains an area of 815 square miles, and used to flow in a shallow winding channel below Preston down to the wide estuary, through which it wandered in a constantly shifting channel, amongst extensive sandbanks, generally branching off into two outlets before reaching the Irish Sea.

Former Condition of the Ribble Estuary. From a series of charts of the estuary prepared at intervals between 1736 and 1836, it appears that there were three outlet channels through which the Ribble in its wanderings flowed into the sea, namely, the north channel following the northern shore past St. Anne's; a middle channel in the centre of the estuary, but varying considerably in position; and the South, or Boghole Channel passing along near the southern shore opposite Southport, which has remained the most stable in position, owing to the main flood tide coming in from the south-west [1] (Plate 9, Fig. 11). With the exception of the Boghole Channel, the position, size, and even the existence of a defined outlet depended upon the course taken by the channel of the river through the estuary In 1736, the main channel of the Ribble went into the South Channel, with only two small shallow channels branching off to a central and a northern outlet; and a similar condition of things existed in 1761. At the date of the next survey, taken in 1820–24, the Ribble followed a fairly

---

[1] 'Report on the Extension of the Navigable Channel of the Ribble to the Sea,' L. F. Vernon-Harcourt, Southport, 1891, pp. 7 to 12; and 'Amélioration de la partie maritime des Fleuves,' L. F. Vernon-Harcourt, Vme Congrès International de Navigation intérieure, Paris, 1892, plate 2, figs. 5 and 6.

direct channel to a central outlet; the northern outlet had entirely disappeared ; and the South Channel, though cut off from the Ribble, still extended nearly 3 miles above Southport, and was the outlet for the Crossens stream draining an area of 55 square miles. The South Channel had become connected with the Ribble again in 1836 ; but the river possessed rather a better outlet by the northern channel, of which no trace existed twelve years previously. Accordingly, the South Channel had proved the most stable and deepest outlet for the Ribble; and in the event of training works being resorted to for fixing and deepening the channel of the river through the estuary, it was clearly indicated by nature as the most favourable outlet for the navigable channel.

Training Works in Ribble Estuary. Up to 1839, the only works which appear to have been carried out on the river were some cross jetties, or groynes, which the landowners placed in front of their property near Preston to prevent the encroachment of the river in its wanderings. About half a mile below Preston, a reef of red sandstone extended across the river, with only a foot of water over it at low water ; and spring tides only rose 6 feet at Preston, whilst neap tides did not get up to the quay. Vessels had to unload their cargoes at Lytham into lighters, drawing only 6 feet of water, for conveyance up to Preston[1].

In 1839 the improvement of the river below Preston was commenced, by blasting and removing the rocky reef ; and the dredging of gravel from the channel was carried out, the materials being deposited at the sides to form training walls on each side of the river. By 1845 the training walls had been extended for 4 miles down the river, and together with the dredging, had increased the range of spring tides to 10 feet at Preston by lowering the low-water line, and had accelerated the flow of the tide up to Preston at springs by one hour ; and vessels of 200 tons, drawing 11 feet of water, were able to get

[1] 'Second Report of the Tidal Harbours Commission,' 1846, p. 10.

up to the quay. The double training walls were subsequently continued down to the Naze; and in 1853 powers were obtained for prolonging the south training wall 4 miles further, and for guiding the river Douglas into the trained channel by training it for two miles down to its confluence with the Ribble (Plate 9, Fig. 11). These works increased the available depth to a maximum of 14 feet at high-water spring tides, and enabled vessels of 300 tons to reach Preston at the highest tides. The channel, however, below the training works continued shallow and unstable ; and the outer part of the south training wall settled, and in one place was cut through by the current, producing a diversion of the channel.

In 1883 the raising of the existing training walls and a further extension of the training works were authorized together with the construction of a dock near Preston and dredging. It was then considered by the engineers of the scheme, that a prolongation of the south training wall for 3½ miles, extending a mile below Lytham, in a slightly con-cave line towards the channel, and the extension of the north training wall for 2 miles, terminating less than half a mile above Lytham, together with the dredging of the river down to 7 miles from the dock, or 2 miles above Lytham, would secure a navigable channel out to sea, with a minimum depth of 6 feet at low-water spring tides, which, with a rise of tide of 26 feet at springs, would give access to the largest class of vessels. It was, however, stated by me in a report to the Corporation of Southport, in March, 1883, that in order to improve effectually the navigation of the Ribble, it would be necessary to prolong the training walls further seawards. The carrying out of the south training wall to the full extent, with an addition of about a mile in length of a mound of dredged material, together with some dredging, had produced by 1890 a general improvement in depth down to below Lytham, and a lowering of three or four feet in the low-water line near Preston ; but a shoal extended across the channel just above Lytham; and from a mile below Lytham, the channel had become worse than in 1883. The depth of channel proposed in 1883 had not nearly been obtained in the trained channel; and in view of the expenditure on the dock and improvement works having considerably exceeded the estimated cost, a Commission appointed by the Board of Trade advised, in 1891, that the dredging should be limited to 5 feet less than contemplated in 1883, which is being carried out[1]. This dredging, when completed, will afford a depth of $23\frac{1}{4}$ feet at high-water spring tides at Preston, and $25\frac{1}{2}$ feet at Lytham, enabling vessels drawing 17 feet to navigate the river between Preston and Lytham at spring tides. In order, however, to obtain a stable channel, and the requisite depth of $26\frac{1}{2}$ feet out to five miles beyond Lytham, it is proposed that the training walls should be extended out on each side of the channel to this distance, which is about two miles short of the bar; but the bar in this instance marks the outer line of the sandbanks rather than the chief shoal at the mouth of the Ribble, which is found about $2\frac{1}{2}$ miles further in. It is, however, admitted that changes in the estuary, resulting from the training works, may necessitate the prolongation of the training works out to the bar, to ensure a stable channel of the requisite depth.

At present the works are confined to the execution of the north training wall above Lytham, authorized in 1883, which has produced a marked improvement in the portion of the channel along which it extends, and the deepening by dredging indicated on the section[2] (Plate 9, Fig. 12). The dredging contemplated in 1883 was commenced at the close of 1886, since which period about 4,173,000 cubic yards have been removed from the channel. The improvement in depth is shown on the section; and the low-water line has been

---

[1] 'Final Report of the Ribble Navigation Commission,' 1891, p. 19.

[2] Some particulars about the present state of the works in the Ribble were given me by Mr. A. F. Fowler, resident engineer to the Ribble Navigation.

lowered about 6 feet at the entrance to Preston dock, augmenting the tidal range at Preston by this amount, and increasing the tidal volume in the channel; whilst the arrival of the flood tide at Preston has been advanced by an hour since 1883.

Effects of the Training Works in Ribble Estuary. The training walls in the estuary of the Ribble have formed a permanent channel in place of a wandering one, and have to some extent increased the depth in the trained channel by scour, which has been further improved by dredging, leading to a considerable improvement in the tidal condition at Preston. The works have, however, led to the reclamation of the whole of the upper part of the estuary, down to the Naze on the northern side, and down to the confluence of the Douglas on the southern side, so that the estuary now commences at the Naze instead of at Preston (Plate 9, Fig. 11). The training of the river, moreover, has enabled the landowners to carry out large reclamations in the southern part of the estuary below the Douglas, which formerly were prevented by the wanderings of the channel; accretion is progressing all along the southern side of the estuary; and the banks to the south of the training works are gradually rising. The completion of the northern training wall also, between the Naze and Lytham, will undoubtedly lead to the accretion of the recess between these places, behind the wall.

Till the commencement of training works on the Ribble, the South, or Boghole Channel was generally the main outlet; and even when cut off for a time from this outlet by the changes in the channel through the estuary, the river soon reverted to this southern outfall, which afforded the best depth, and was the most favourable for the admission of the flood tide. The direction, however, given to the trained channel, as close as practicable to the northern shore, tended to divert the river from its southern outlet; but even in 1850, a narrow subsidiary channel of the river communicated with

the South Channel. Soon after, however, the prolongation
of the training walls near the northern shore, the training of
the Douglas into this northern channel, and the resulting
accretion and reclamation on the southern side of the estuary,
diverted the river permanently from the Boghole. The influence of these works was already apparent in the chart of
1860, which exhibits the channel of the river close along the
northern shore past Lytham, and then dividing into a northern
and central outlet, a continuous unbroken stretch of sand
between the southern shore and the channel in the estuary
above Crossens, and the South Channel cut off from the
main channel except at half tide, and connected only with
Crossens stream [1]. By degrees the effect of the training works
on the channel and the resulting accretion and reclamations
produced a raising of the intervening sandbanks, and caused
a considerable narrowing and curtailment of the South
Channel, and by 1883 had cut it off from Crossens stream
which afforded its sole remaining means of maintenance [2].
The raising of the old training walls since 1883 and the
prolongation of the south training wall have still further
promoted accretion in the southern portion of the estuary,
and the raising of the banks between the trained channel and
the Boghole. Eventually, unless Crossens stream is brought
back into the Boghole, this South Channel, which afforded so
good an outlet for the Ribble in former times, must disappear,
and thus remove from Southport the only appearance of sea
which it possesses at low water.

Three effects of training works are clearly manifested by
the results of training the Ribble, namely, the improvement
and stability of the trained channel, the inefficacy of training
walls to improve the channel beyond their termination, and

---

[1] 'Report on the Extension of the Navigable Channel of the Ribble to the Sea,'
L. F. Vernon-Harcourt, Southport, 1891, pp. 9 to 14.

[2] 'Report on the Crossens Channel New Cut,' L. F. Vernon-Harcourt, Southport, 1884.

the occurrence of accretion in the portions of a sandy estuary severed by the training walls from the main channel. The South Channel has already been effectually cut off by the training works from the Ribble, and is in course of silting up ; and further extensions of the training works, needed for providing a navigable outlet, will hasten this process. The northern channel, which is far less stable than the South Channel, has not as yet been affected by the training works ; but it too will eventually be cut off permanently from the Ribble by the extension of the north training wall past Lytham, and will disappear.

Remark son Ribble Training Works. There were originally three outlets available for the channel of the Ribble through its estuary ; and though under ordinary conditions, the direct central outlet would appear the most natural, a study of the charts would have revealed that the southern outlet afforded the most stable and deepest channel. The indications, however, of nature were overlooked ; and the channel was trained in a narrow channel near the northern shore of the estuary. If Southport had been a seaport, its interests could have hardly been so disregarded as in the course chosen for the channel in 1853 ; and at that period Southport had not developed into the important place it has since become ; and its inhabitants were probably unaware of the effect the training works proposed would have upon its welfare. The extension of the training works since 1883 was the natural sequence of the works of 1853 ; but this, and the further prolongations proposed, if carried out, will by degrees complete the deterioration of the South Channel. The training of one channel in a sandy estuary subject to accretion, necessarily involves the disappearance of the other channels. Assuming that, in spite of the indications of the early charts, the central channel was the most suitable for Preston, it is evident that the trained channel should have been widened out more seawards, even though this would have involved a larger amount of dredging; for as the accretions consequent on the training works progress, the estuary will become very much restricted, and a wider trained channel would have ensured a greater tidal scour for maintaining the outlet. The engineers, indeed, from the outset deprecated reclamations, and merely aimed at improving the navigable channel : but unfortunately in an estuary like the Ribble, with an abundance of sand outside, it is impossible to carry out training works without promoting accretion at the sides of the trained channel, which involves the loss of tidal scour at the outlet and favours reclamation. This inevitable result was undoubtedly hastened in the case of the Ribble by constructing half-tide training walls, and raising them by degrees; but probably if the effect of training works had been duly appreciated, a wider and more central channel would have been designed, and dredging to provide the necessary deepening would have been more largely resorted to.

LOWER AND OUTER WESER TRAINING WORKS.

The river Weser drains an area of 18,600 square miles; and its discharge varies between 196 cubic yards per second at a low stage, and a maximum of 5,200 cubic yards during a great flood, its average discharge being 387 cubic yards per second. Formerly the tide did not flow higher up than Bremen, 42 miles above Bremerhaven, which is situated at the head of the actual estuary, where the rise of tide averages $10\frac{4}{5}$ feet [1] (Plate 7, Figs. 3 and 4). About 10 miles below Bremerhaven, the Weser estuary joins the estuaries of the Jahde and the Elbe on either side, through which these rivers flow in separate channels into the North Sea, which the Weser reaches about 36 miles below Bremerhaven. At Heligoland in the North Sea, beyond the combined estuaries of these

[1] 'L'Amélioration des Fleuves dans leur partie maritime,' L. Franzius, Vme Congrès International de Navigation intérieure, Paris, 1892.

rivers, the rise of the tide is only 6 feet, so that the range of tide is increased nearly 5 feet in flowing up to Bremerhaven.

Training Works in the Lower Weser. The river between Bremen and Bremerhaven flows through flat alluvial land, which may have originally formed an extension of the estuary. The floods of the river and the changes of the channel have been restricted by embankments on each side, at varying distances back from the river. The strips of land, however, thus enclosed with the river, though having about three times the width of the river in some places, appear to serve only as receptacles for floods, and are at too high a level to be of value as reservoirs for tidal water to increase the tidal capacity of the river. The training works, accordingly, along this portion of the river, carried out since 1887, resemble the regulation works on the Maas below Rotterdam and the Nervion below Bilbao, rather than training works through an estuary. They possess, however, a special intrest in systematically training both a high-water and a low-water channel, by double rows of fascine mattresses on each side, with enlarging widths seawards carefully calculated to provide an adequate channel for the full tidal flow as well as the fresh-water discharge (Plate 9, Figs. 8 and 10). The course of the river has been straightened, bends have been eased, and secondary channels suppressed, so as to facilitate the tidal flow and ebb, and to concentrate the flow in a single channel; whilst dredging has been employed to increase the improved depth obtained by scour (Plate 9, Fig. 9). The width of the low-water channel has been adjusted, so that whilst large enough to give free passage to the river from a little before to a little after low water, it may not allow the low-water channel to meander between the training walls. The material dredged from the trained channel is deposited at the back of the training walls, and is also used to fill up the abandoned secondary channels. The works involve the removal of about 72 million cubic yards, of which it is estimated that $40\frac{1}{2}$ million cubic yards will have to be dredged or excavated, the remainder being removed by scour.

Training Works in Weser Estuary. The training works were prolonged into the estuary below Bremerhaven in 1891, where a submerged shoal exists in mid-channel in front of the outlet, dividing the main channel into two branches. A single training wall, $3\frac{3}{4}$ miles long. in continuation of the left bank of the river, provides an enlarging outlet for the river, which is bounded by the shore-line on the other side (Plate 9, Fig. 8). This training wall skirts the edge of a large sandbank, and is to be prolonged seawards to such an extent as may be requisite to secure an adequate depth in the outlet channel. This portion of the channel is also to be dredged to the depth indicated on the section (Plate 9, Fig. 9).

Results of Weser Training Works. The training works aided by dredging increased the available navigable depth at high water up to Bremen, in the first four years, from 9 feet to $14\frac{3}{4}$ feet; and this depth will gradually be augmented to about 22 feet by dredging (Plate 9, Fig. 9). The training wall in the estuary, within six months after its completion, had increased the average depth over the bar by $1\frac{3}{8}$ feet, and $4\frac{1}{4}$ feet in the deepest channel.

The rectification and deepening of the channel has accelerated the passage of the flood tide up the river, and has extended the tidal limit from Bremen up to Habenhausen, $4\frac{1}{4}$ miles higher up (Plate 7, Fig. 3). The low-water line had already been lowered materially in 1891 between Bremerhaven and Bremen, reaching a maximum of about $3\frac{1}{3}$ feet at Vegesack (Plate 9, Fig. 9); and a further lowering may be confidently anticipated from the proposed deepening by dredging. The volume of tidal water passing Bremerhaven had been increased by more than one-sixth in 1891, and the efficiency of the ebb correspondingly augmented.

Remarks on Weser Training Works. The training of

a low-water, as well as a high-water channel, and the employment of dredging to obtain the necessary depth, have enabled
any material contraction of the high-water channel to be
avoided. This maintenance of the width of the high-water
channel, combined with the lowering of the low-water line by
the improvement of the river, has produced an increase in the
tidal capacity above Bremerhaven, in spite of the shortening of
the channel by its rectification and the filling up of secondary
channels. The training works in the estuary below Bremerhaven have, accordingly, been commenced under the favourable
condition of an increased tidal flow. Moreover, the outlet is
being given a trumpet shape, so as not to impede the influx
of the flood tide up the river; and as fairly deep water
approaches within 9 miles of Bremerhaven, it may be anticipated that by a prolongation of the training wall in a
diverging line from the opposite shore, and the assistance of
dredging, a deep approach will be provided for the Weser,
and that sea-going vessels of large draught will have access
to Bremen which has already been converted by the training
works into a seaport.

CHAPTER XIII.

TRAINING WORKS IN ESTUARIES (*continued*).

Physical Conditions of the Loire and the Seine contrasted. *Loire Estuary Works:* Description of Estuary; Training Works below Nantes, their extension and results; Ship-Canal formed alongside part of Estuary to provide a better navigable depth; Dredging in Estuary, and in Trained Channel; Improvement of Navigable Depth up to Nantes; Remarks on Loire Training Works, difficulties involved *Seine Estuary Works:* Original state of Tidal Seine; Training Works in Estuary, their extent and consolidation; Effect of Training Works on Navigable Channel; Accretions in Seine Estuary, their origin, extent, and volume; Alterations in Lines of Depths at Outlet; Tancarville Canal, description and object of Canal; Schemes for Completion of Seine Training Works, difficulties, variety of proposals, their aims and merits; Remarks on Seine Training Works, value for Rouen, cause of anxiety for Havre; Importance of Completion of Works, in trumpet-shaped form aided by dredging. *Remarks on Training Works:* General Conclusions deduced from Experience; Enlargement of Trained Channel, indicated by instances. Concluding Remarks.

Two rivers of France, the Loire and the Seine, presenting considerable differences in their physical conditions, have been partially trained through their extensive estuaries, for the purpose of improving the navigable access to the ports of Nantes and Rouen respectively. The Loire is a torrential river, and brings down large quantities of detritus; whilst the Seine is comparatively gently-flowing, and its fresh-water discharge is remarkably free from alluvium. The estuary of the Loire, moreover, is somewhat narrow, and has a contracted outlet; whilst the estuary of the Seine is much wider, and has a large outlet; but though free from the islands which are found in the upper estuary of the Loire, the Seine estuary is much more encumbered by sandbanks (Plate 9, Figs. 1 and 4). The Loire drains an area of 45,000 square miles, about half as large again as the Seine basin with an area of 30,370 square miles; but the tidal rise at its mouth is only 16½ feet at springs, as compared with 23½ feet at the mouth of the Seine; whilst, owing to this smaller tidal range, in conjunction with the greater slope of the country through which it flows, the Loire is tidal for only about 44 miles from the sea, whereas the Seine is tidal up to Martot, about 90 miles inland.

The discharge of the Loire is very irregular, falling to a minimum of 160 cubic yards per second in a very dry season, and attaining 7,600 cubic yards per second in exceptional floods, or more than double the maximum discharge of the Seine. This small summer flow renders the Loire very shallow in dry weather, and therefore unsuitable for regular navigation; and the traffic on the river above Nantes is very small. The Seine, on the contrary, has a much more regular flow; and its small fall between its tidal limit and Paris, has rendered it capable of being greatly improved in navigable depth by canalization. Rouen, accordingly, being on the way to Paris, and connected with it by a very accessible waterway, serves as the port of transit between the sea and an important inland waterway leading direct to the capital; and its prosperity depends on its access to the sea. Nantes being devoid of inland communication by water, and not being on the route to any important town, is a port of considerably minor importance; and its trade is due to its being near the head of the tidal Loire, and depends on its connexion with the sea. Both these river ports, however, are somewhat overshadowed by the sea-ports of Havre and St. Nazaire. which are situated at the entrance of the Seine and of the Loire respectively. The commercial value, indeed, of an inland position at the head of a tidal navigation is shown by these inland ports not having been eclipsed by their formidable rivals on the sea-coast, and by their trade flourishing in proportion to the improvement of their communication with the sea.

X

LOIRE ESTUARY WORKS.

Nantes is situated near the head of the Loire estuary, about 35 miles above its outlet, and 9½ miles below Mauves, the limit of the tidal flow at a low stage of the river. The tidal rise of 16½ feet at the mouth of the estuary opposite St. Nazaire, is reduced to 6 feet at Nantes (Plate 9, Figs. 4 and 5); but during ordinary floods the tide is not felt within 12 miles of Nantes, being overpowered by the flood discharge; and in very high floods the flood-tide current is imperceptible, except near the bed of the river at St. Nazaire. The estuary between Nantes and La Martinière consisted of a network of channels flowing between numerous islands, probably a gradual growth from the ordinary state of an estuary encumbered with sandbanks, resulting mainly from the deposit of material brought down by the river. Islands are also found in the more open estuary below La Martinière, and nearly down to Paimbœuf; but the alluvium in this part of the estuary above low water was formerly of marine origin.

Training Works in Upper Estuary of the Loire. Proposals to improve the access to Nantes were made in the seventeenth century; but the first works were carried out in 1756–68, at which period vessels of 8 feet draught had difficulty in getting up to Nantes[1]. These works consisted in barring the secondary channels between the numerous islands and the left bank of the estuary, by means of submerged rubble-stone dams, from Bouguenais down to Couëron, so as to concentrate the main current into the channel bordering the right bank, with the view of increasing its depth by scour[2]. Groynes also were carried out from the right bank into the channel, in order to contract the waterway still more, and thereby

---

[1] 'Annales des Ponts et Chaussées, 1878 (2), p. 563.
[2] 'Ports Maritimes de la France,' vol. v, p. 257.

increase the scour. Considerable accretions, however, resulted from these works; and no material improvement in the depth of the navigable channel was effected.

No further works were undertaken till 1834, when longitudinal training walls, raised to a level of $3\frac{1}{4}$ feet below highwater spring tides, were constructed in different places between Nantes and Le Pellerin, mainly on one side of the channel, so as to narrow it to a width of 820 to 1,000 feet. Even these works, however, completed in 1840, produced merely a slight improvement in the channel, for in 1850 the depth over the principal shoals had only been increased between 8 inches and 1 foot.

In 1859–65 training works were again carried out along the ten miles between Nantes and La Martinière. Rubblestone training walls, pitched on their exposed faces, were formed along each side of the channel, with their tops at the level of high water of low springs, placed 650 feet apart near Nantes, and widening out gradually to 1,000 feet at Le Pellerin (Plate 9, Figs. 4 and 7). A gradual increase in depth was effected by these works, amounting in 1876 to an average of 5 feet over the whole distance, and $3\frac{1}{4}$ feet across the worst shoals, giving an average depth in the navigable channel of 20 feet, and a minimum of $14\frac{3}{4}$ feet at high-water spring tides[1]. The shoals were found in places where the old training walls had not been reconstructed, or where side channels were left open for the convenience of riparian owners; and they might have been removed by the completion of the training walls where the old walls were left unaltered, and across the open gaps.

The improvement in depth of the trained channel was accompanied by a large extension of the sandbanks in the estuary below, between La Martinière and Paimbœuf, so that this

---

[1] A plan and sections of the Loire were furnished me by Mr. Lefort, engineer-in-chief of the tidal Loire, and also a summary of the works carried out, and their results.

section of the channel, which previously was deeper than the portion above, became worse. This deterioration was due to the deposit in this part of the material scoured out of the trained channel, to the loss of tidal scour by the large accretions which had taken place at the back of the training walls, and to the direct discharge of the detritus of the river into this section of the estuary. The mean depth of the navigable channel through this part of the estuary was reduced from 20½ feet to 18 feet between 1859 and 1876; and the length of the shoals with a less depth than 17⅓ feet over them, was doubled in the same period; whilst fresh islands have been formed, or are in course of formation.

Ship-Canal alongside the Loire Estuary. In order to provide the requisite depth in the deteriorated portion of the estuary below La Martinière, the prolongation of the training walls was proposed; but eventually, in 1879, the alternative scheme of a lateral ship-canal was adopted, so as to avoid the shoaling circuitous channel through the estuary below the existing training works. This ship-canal has been constructed through the land on the left bank, alongside the estuary, starting at La Martinière from the trained channel, and after a course of 9⅓ miles, rejoining the estuary at Carnet, 4⅓ miles above Paimbœuf, where there is a deep channel close to the shore, passing between Carnet Island and the left bank of the estuary (Plate 9, Fig. 4). The canal was completed in 1892; it has a bottom width of 78¾ feet, and a depth of water of 19⅔ feet; and it has a lock at each end, 558 feet long and 59 feet wide.

Dredging in the Loire Estuary. Since 1877, dredging has been carried out in the estuary on a much larger scale than previously; and up to the end of 1892 it was mainly executed in the shallow part of the estuary, between La Martinière and Paimbœuf, increasing the available depth in the navigable channel through this section of the estuary from 13 feet to 15¾ feet at high-water spring tides. The

trained channel between Nantes and the entrance to the
ship-canal, is being deepened by dredging to a depth of
21 feet below high-water spring tides; and the channel
between the outlet of the canal and deep water in the estuary
is being deepened to 24 feet. Dredging will still be carried
on to a moderate extent in the portion of the estuary super-
seded by the ship-canal for large vessels, in order that the
flood tide may not be impeded in its passage up the estuary
by shoaling in the channel, and that small vessels may continue
to go up to Nantes without being obliged to pass through
the ship-canal.

The ship-canal, and the increase in depth above and below
effected by dredging, will enable vessels drawing 16½ feet to
reach Nantes from the sea in a single tide, even during the
lowest neaps, which formerly was only practicable on at
most fifty days in the year; and vessels of 19 feet draught
will be able to accomplish the passage during spring tides
and in flood-time.

Remarks on Loire Training Works. The estuary of the
Loire is not favourably circumstanced for improvement by
training works, on account of the large quantities of sand
brought down by the floods of the river, which, together
with alluvium brought in by the sea, readily deposit at
slack tide in sheltered parts of the estuary. The training
walls caused the river detritus to be discharged into the
portion of the estuary beyond their termination, and at the
same time reduced the tidal scour in this part by the diminution of tidal capacity produced by accretion in the estuary
behind the walls. Accordingly, the training works terminating
at La Martinière merely shifted the position of the shallowest
channel lower down the estuary; and the deteriorated portion
of the estuary, between La Martinière and Paimbœuf, has
ceased to be used as the navigable channel since the construction of the lateral ship-canal. In spite, however, of its abandon-
ment for navigation, this portion of the estuary will have to be

maintained by dredging, lest a reduction in tidal scour should injure the lower estuary and the approach to St. Nazaire.

The enlargement in width between the existing training walls as they descend towards the sea, has not been made adequate, owing doubtless to the desire to effect the requisite deepening of the channel by scour. Moreover, in order to provide a navigable channel of sufficient depth through the estuary between Nantes and the sea, it would have been essential to continue the training walls down to Paimbœuf, gradually enlarging the trained channel to the full width of the narrowed estuary, and securing the required depth by dredging. Nevertheless, a trained channel, however well designed, would inevitably restrict the reservoir available for the materials brought down by the river, and reduce the tidal capacity of the estuary, and consequently diminish the tidal scour past St. Nazaire and over the bar outside. Nantes would have benefited by such works, for the depth in its improved approach channel is less than the available depth over the bar. With a large port, however, at the outlet of the estuary like St. Nazaire, it would be inexpedient to carry out works in the estuary which would reduce the tidal capacity of the estuary, which serves as a sort of natural scouring basin for maintaining the depth of the channel over the bar, upon which a deep approach channel to St. Nazaire depends.

SEINE ESTUARY WORKS.

The river Seine flows from Rouen down to La Mailleraye, a distance of 39 miles, in a stable bed, through rock, clay, and gravel; and throughout this portion of its course, which is about half the distance between Rouen and the sea, it possesses a good natural depth, so that by merely dredging a few shoals near Rouen, a depth of 24 feet at high-water neap tides has been provided. Below La Mailleraye, however, where the estuary may be said to have commenced, the river formerly flowed to the sea in a shallow, winding, and variable channel through shifting sandbanks. In some places within twelve miles of La Mailleraye, the shoals were so high that the available depth over them at high-water spring tides was only 10 feet; vessels exceeding 200 tons could not ascend the river; and vessels of between 100 and 200 tons often had to wait for a high enough tide. The voyage between the sea and Rouen never occupied less than four days; and the perils of navigating the shallow, shifting channel were aggravated by the bore, which travelled up the river with considerable velocity, as a breaking wave some feet in height, at the commencement of the flood-tide during springs; and consequently wrecks frequently occurred. Various schemes were proposed during the eighteenth century for remedying this disastrous condition of the estuary, as for instance a ship-canal along the left bank down to Honfleur in place of the estuary channel, and the improvement of the channel through the estuary by weirs, groynes, or training works of various kinds[1].

Training Works in Seine Estuary. The first works undertaken for the improvement of the navigable channel through the estuary were commenced in 1848, by the construction of training walls from Villequier, 7 miles below La Mailleraye, down to Quillebeuf, a distance of about 11 miles. These training works, completed in 1851, formed an experimental portion of a general scheme, proposed in 1845 by Mr. Bouniceau, for training the channel from La Mailleraye to Honfleur, and across the estuary to Havre[2]. As the depth over the shoals was rapidly increased by these works from $10\frac{1}{2}$ feet to $21\frac{1}{8}$ feet, it was determined to extend them up to La Mailleraye, and down to Tancarville on the northern side, and to La Roque on the southern side, which extensions were completed in 1859 (Plate 9, Fig. 1). The training walls were placed

[1] 'The River Seine,' L. F. Vernon-Harcourt, Minutes of Proceedings Institution C. E., vol. lxxxiv, pp. 241–252, and plate 4.

[2] 'Étude sur la navigation des rivières à marées,' P. Bouniceau, 1845, p. 152.

980 feet apart at Villequier, widening out to 1,310 feet at Quillebeuf, and 1,640 feet at Tancarville; and they provided a deep stable channel down to Tancarville. Below this point, however, the channel diverged northwards, and was shallow, circuitous, and changeable, so that a prolongation of the walls down to Berville was carried out, which was completed in 1869. A further extension of the works, proposed in 1871, was relinquished in consequence of the opposition of Havre, which had become alarmed by the changes in the estuary produced by the training works already carried out; so that the training walls have been stopped in the middle of the estuary, at the point they had reached at that period.

The training walls, originally formed of mounds of rubble chalk obtained from the neighbouring cliffs adjoining the river in places, were raised to high-water level down to Tancarville on the north, and to La Roque on the south; but below these points, they were only carried up two or three feet above low-water spring tides, with the object of avoiding as far as possible further accretions in the estuary. The high training walls, being eroded by the bore, and undermined by the deepening of the channel, were disturbed, and settled; and after being raised by additions of rubble chalk for several years, they were eventually protected against the wash of the bore and undermining, by a concrete facing and apron, and by piling and planking along the toe (Plate 9, Fig. 3). The low training walls, moreover, below Tancarville and La Roque were so much injured by the rush of the tide across them at certain periods, that they also have had to be reconstructed and raised to mean high-water level.

Effects of Training Works on Channel. A comparison of the longitudinal section of the river in 1824 with the sections along the trained channel in 1880 and 1891, shows the remarkable increase in depth that has been effected by the training walls between La Mailleraye and Berville (Plate 9, Fig. 2). The deepening averages about 15 feet, the minimum reaching 6 feet; and the minimum navigable depth between Havre and Rouen at high-water neap tides has been increased from 10 feet up to 18 feet. The low-water line also has been lowered throughout the trained channel, and above, by the increased depth, the lowering reaching a maximum of about 8 feet at Tancarville. The improvement was almost wholly due to the training walls, for the only dredging effected in the trained channel was the lowering of the hard Meules bank about a mile and a half below La Mailleraye. The arrival of the flood tide at Rouen has been advanced one hour by the removal of the obstacles to its progress above Berville; but the bore is still experienced along the comparatively narrow trained channel, reaching a maximum at Caudebec, and disappearing at Rouen (Plate 7, Fig. 1).

The influence of the training walls on the channel extends very little beyond their extremities. Fortunately as regards the depth, the worst shoals were above Berville; but on emerging from between the training walls, the channel remains circuitous and variable along the 11 miles of sandbanks separating Berville from deep water outside the estuary. The training works, however, have brought deep water in the river much nearer the open sea, so that the intervening length of shallow channel is readily traversed at high water in fine weather. Moreover, owing to the greater range of tide towards the mouth of the estuary, the depth near high water in the main channel approximates to the available depth in the trained channel; so that little inconvenience would be experienced by the absence of an improved depth between Berville and the sea, if the channel was stable and direct. Though, however, the winding and shifting channel is marked by buoys, it is difficult for vessels of large draught to navigate the intricate channel during storms, when alterations are most liable to occur, and in fogs. Accordingly, in order to render the passage through the estuary always practicable and safe at high tide, for the sake of Rouen which has been raised by the training works to the position of the fifth seaport of France, it will be necessary to prolong the training walls.

Accretions in the Seine Estuary. About half of the deeply recessed portion of the estuary lying between Quillebeuf and La Roque, comprising mainly the Vernier marsh, had been reclaimed previously to the commencement of the training works ; and a smaller portion of the foreshore of the recess on the northern side above Tancarville had also been enclosed. The rest of the estuary, however, was covered by the tide, which came up close to the base of the chalk cliffs which bounded the estuary on the north between Tancarville and Harfleur, and on the south between La Roque and Honfleur (Plate 9, Fig. 1). In proposing his scheme of training works, Mr. Bouniceau expressed the opinion that the accretion that might result from them would be exceedingly slow in accumulating, and that the filling up of the estuary behind the training walls would extend over what might be regarded as a geological period, amounting to 210 centuries. He considered also that the deposit which might settle behind the training walls would be derived in great measure from the sandbanks further down the estuary, which would thereby be lowered, constituting in fact an advantageous transposition of material.

Owing to the freedom of the Seine waters from sediment, the estuary is practically in no danger of accretion from this source. The flood tide, on the contrary, enters the estuary laden with material eroded by waves from the Calvados cliffs, and collected by the tide in passing along this extensive stretch of coast, especially when storms stir up the alluvium composing the foreshore. Formerly, however, the alluvium brought in by the flood was carried out by the ebb, especially when the banks were periodically stirred up by the shiftings of the channel. The sands encumbering the estuary were doubtless originally derived from the Calvados coast; whilst the cliffs to the north have supplied the shingle beach in front of Havre, which extends inside the estuary to Hoc Point. No records are available to indicate whether accretion was taking place in the untrained estuary; but it is clear that the process was slow, except in sheltered recesses. As soon, however, as the existing state of equilibrium of the estuary was disturbed by the construction of training walls, which concentrated the main tidal flow and ebb and the whole of the fresh-water discharge into the trained channel, the settlement of alluvium brought in by the flood tide was promoted in the slack water, under shelter of the training walls and out of the run of the currents; and accretion rapidly took place. The silting up of the large recesses above La Roque, already partially reclaimed, was undoubtedly accelerated by bringing the training walls close against the projecting points at Quillebeuf and Tancarville, and constructing a cross wall from La Roque point to the south wall, thereby enclosing these areas which have since been reclaimed (Plate 9, Fig. 1). The accretion, however, resulting from the training works has not been confined to these enclosed portions of the estuary, nor even to the sides of the estuary at the back of the training walls, but has crept down along the sides of the estuary to Honfleur on the south side, and to Harfleur on the north side, 6¼ and 9 miles respectively below the ends of the training walls. Accordingly, though the training works lose their effect in deepening and fixing the channel close to their extremities, their influence on the accretion of the estuary extends several miles lower down.

A survey of the estuary made in 1875 showed that within thirty years of the commencement of the training walls, after making allowance for the increase in depth of the trained channel, the capacity of the estuary at high tide had been diminished by 272 million cubic yards, owing to the accretion resulting from the works[1]. The rate of accretion would

---

[1] 'Reconnaissance hydrographique de 1875 à l'Embouchure de la Seine.' Report by X. Estignard, 1877.

naturally attain a maximum about the period of the completion of the training works; and subsequently the estuary
would gradually approach again a position of equilibrium,
as soon as the accretion induced by the works has taken
place. Another survey made in 1880 showed that in the five
years that had elapsed since the previous survey, a further
reduction in the capacity of the estuary had occurred, to the
extent of 40 million cubic yards, raising the total accretion
in the estuary since the commencement of the works to
312 million cubic yards, more than a third of the amount
to the accumulation of which Mr. Bouniceau had allotted an
almost indefinite period. Accretion was evidently still progressing in the estuary in 1880, though at a diminished rate.
Without a fresh survey, it would be impossible to determine
the amount of the subsequent accretion; but a comparison of
the low-water charts of 1880[1] and 1891 (Plate 9, Fig. 1) shows
that the area of marsh lands has somewhat increased on the
north side, that the area of sandbanks above low water has
become greater, and that more sandbanks emerge near the
outlet. It appears therefore that the accretions due to the
training works have not yet reached their limit. The time of
high water at Havre has advanced about forty minutes since
the construction of the training works, which is evidently due
to the quicker filling of the estuary on account of its greatly
reduced tidal capacity.

The idea entertained by Mr. Bouniceau that the material
deposited at the back of the training walls would come from
the sandbanks at the mouth of the estuary, was at first so far
realized that the 5- and 10-metre lines of soundings advanced
into the estuary, after the commencement of the works, up to
1875. Since that period, however, they have been receding;
and in the chart of 1891, they were 3 and $3\frac{1}{3}$ miles seaward of
their positions in 1875, and further out than in any previous
chart back to 1834. This advance and retreat of the lines of

---

[1] Minutes of Proceedings Institution C. E., vol. lxxxiv, plate 4, fig. 1.

soundings is readily explained. The sand brought up from the mouth, as well as from outside, by the flood tide, found a resting-place in the sheltered parts of the estuary during the extension of the training walls, and for some time after their completion, and therefore did not return on the ebb to be deposited again; and consequently the depth at the outlet improved. As soon, however, as the main part of the accretions resulting from the training walls had taken place, the reservoir for the sand brought up by the flood tide having been filled up, the sand returned with the ebb tide; and the scour of the ebb having been greatly reduced by the diminution in the tidal capacity of the estuary, the sand was more readily deposited, and reduced the depth at the outlet. This will continue till the cessation of accretion, and the reduction in the section of the outlet channel, corresponding to the diminished tidal scour, produce a fresh state of equilibrium in the estuary.

Tancarville Canal alongside Seine Estuary. The idea of avoiding the dangers of the passage of the estuary, by connecting Havre with the river above by a canal, appears to have originated in the eighteenth century; and various projects have been brought forward from time to time. At length in 1876, at the request of the authorities of Havre, the question was investigated; the construction of a canal connecting the docks at Havre with the river Seine at Tancarville was authorized in 1880; and the canal was opened for traffic in 1887 [1]. The canal which traverses the alluvial deposits between Harfleur and Tancarville, has a lock at each end, to regulate the water-level which is kept below the highest tides, and an enlargement in width near each end to serve as a siding (Plate 9, Fig. 1). The present object of the canal is to enable river craft to pass between Havre and the river without traversing the estuary; and therefore the canal has only been excavated to a depth of $11\frac{1}{2}$ feet between Tancarville and Harfleur. The locks, however, with an entrance width of $52\frac{1}{2}$ feet, have

---

[1] 'Canal du Havre à Tancarville,' Maurice Widmer, Havre, 1887.

had their sills placed low enough to afford a depth of water over them of 23 feet; and their width, and the width of the canal have been made of such a size that it will be only necessary to deepen the canal to convert it into a ship-canal, capable of accommodating sea-going vessels, in the event of the estuary becoming so much deteriorated by accretion as to be no longer suitable for navigation. As the communication of Harfleur with the sea has been cut off by the accretions in front of it, and by the canal, a branch canal has been carried up to Harfleur; and the depth of the waterway between Harfleur and Havre has been made $19\frac{3}{4}$ feet (Plate 11, Fig. 22). The canal is $15\frac{1}{2}$ miles long; and its cost, including a basin at Havre, quays, nine swing-bridges across the locks and for roads, and land, amounted to £928,000. The traffic through the canal reached 287,500 tons in 1893.

Schemes for Prolongation of Seine Training Works. The training walls which were stopped at Berville in 1869, form part of an incomplete scheme; whilst the provision made for enabling the Tancarville Canal to be readily converted into a ship-canal, indicates the uncertainty which exists in some quarters as to the maintenance of the navigable capabilities of the estuary. It would be impossible to obtain a satisfactory and stable navigable channel between Berville and the sea without a prolongation of the training walls; whilst any extension of the training walls will introduce a fresh disturbance in the state of equilibrium to which the estuary is at present approaching. Moreover, any scheme must secure Honfleur its access to the sea, without in any way endangering the maintenance of the depth of the approach channel to Havre. The existence of ports on opposite sides of the estuary occasions a difficulty in forming a trained channel convenient for both; and it would be quite inadmissible to prolong the trained channel to Honfleur, and then to carry it across the estuary to Havre, as indicated in one of the original schemes, for a channel trained across the estuary could not be maintained with low training walls, and with high training walls would lead to the immediate reclamation of the estuary.

Various schemes have been proposed for the completion of the Seine training works, presenting a great diversity in design[1]; but they may be grouped under three distinct classes. One set of schemes proposes to close three-fourths of the outlet of the estuary, by a breakwater extending from Villerville to the Amfard bank ; and the prolongation of the south training wall in a continuous curve past Honfleur towards Havre, would guide the channel to the narrow outlet (Plate 10, Figs. 1 and 2). In a second series of designs, the channel would be trained in a sinuous course between two training walls, widening out seawards, the southern wall terminating at Honfleur, and the northern wall being either stopped in mid-estuary or prolonged to Havre (Plate 10, Figs. 4 and 5). The third class of schemes would train the channel in an expanding trumpet-shaped form to the outlet (Plate 10, Figs. 3 and 6). The first class of designs would convert the estuary into a huge sluicing basin, and thereby secure a deep channel through the narrow outlet: the second aims at forming a deep and stable channel by a succession of curves ; and the third would endeavour to guide the channel in a central course, whilst freely admitting the flood tide, and reducing the capacity of the estuary as little as practicable consistently with the guidance of the channel. A great reduction in the width of the outlet, whilst improving the depth in the narrowed channel, would check the influx of the tide, and consequently reduce the tidal rise in the enclosed estuary, involving a diminution in tidal capacity, and a loss of depth in the navigable channel at high water. A sinuous course, though producing a fairly stable channel, especially at the concave bends, in a river where the flow is always in one direction, does not afford a similar advantage in a tidal river, owing to the divergence in the courses of the flood and ebb

[1] Minutes of Proceedings Institution C. E., vol. lxxxiv, plate 4, fig. 9.

tides in a wide curving channel; and any scheme where a training wall terminates in mid-estuary must be regarded as incomplete (Plate 10, Fig. 5). A trumpet-shaped outlet for the trained channel is the only form which follows the indications of nature, though some uncertainty exists as to the best rate of enlargement, and some difficulty would be experienced in combining a wide outlet with an adequate deepening of the channel. The rate of enlargement, however, is capable of calculation for any special case; and deepening by dredging might be advantageously resorted to within the trained channel, aided possibly by inner low-water training walls or dipping cross dykes, in order to avoid an undue restriction of the estuary.

Remarks on Seine Training Works. The training works in the Seine estuary are remarkable for the great improvement in depth effected by them as far as they have been carried, the large and unexpected accretions which they have occasioned, as well as the distance down the estuary to which these accretions extend, and the great diversity of opinions and schemes to which the proposed completion of the training works has given rise. Rouen has gained enormously by the great increase in depth effected in the worst part of its navigable access to the sea, lying between La Mailleraye and Berville; though the portion of the channel between Berville and the sea still needs amelioration to satisfy the requirements of this flourishing inland port. Anxiety has, at the same time, naturally been aroused about the maintenance of the deep-water access to Havre, in consequence of the large deposits which have accumulated in the estuary, creeping down to the neighbourhood of Havre; and the retrogression of the lines of soundings seawards since 1875 is not calculated to allay these fears. It has, accordingly, been proposed that the completion of the training works should be accompanied by the formation of a large outer harbour on the sea-coast in front of Havre, so as to provide the port with a new deep sheltered outlet beyond the influence of the deposits in the Seine estuary.

The completion of the training works should not be longer delayed in the interests of Rouen and Honfleur, which at present are exposed to the inconveniences of a constantly shifting outlet channel. The training walls have hitherto been placed too close together to ensure adequately the free admission of the flood tide, and to provide for the due enlargement of the outlet. The trained channel was made narrow in order to increase the deepening by scour; but it would have been much better to give the channel a greater rate of enlargement, and to have gained the requisite depth by the aid of dredging as well as scour. The narrowness of the existing trained channel has proved a hindrance in the designs for the completion of the works; and if possible the widening out should be commenced at Quillebeuf, so as to render a suitable expansion to a wide outlet more practicable; for the northern training wall must not impede at all the access to Havre, and the southern training wall should not project from the coast so as to cause an advance of the foreshore in front of Trouville (Plate 10, Fig. 1). A trumpet-shaped outlet would offer the least interference with the existing condition of the estuary and the depths outside, and the best prospect of obtaining a fairly stable central channel, which could be deepened and fixed by dredging and low inner training works. Moreover, by widening the trained channel above Berville, and regulating and deepening the outlet channel below, the influx of the flood tide would be facilitated, and the bore consequently reduced.

Even under existing conditions, vessels entering the estuary a little before high water can almost always get up to Rouen in a single tide, owing to the long duration of high water; but the removal of some shoals in the trained channel would render the passage up the river still easier. The descent of the river from Rouen to the sea can only be effected in two,

Y

or occasionally in three, tides ; and the vessels have to anchor in one of the deep pools on their route during low tide. The lowering of the shoals, however, in the trained channel, as proposed, will enable vessels drawing less than $16\frac{1}{2}$ feet to get down without a stop ; and vessels of larger draught will never have to anchor more than once on the journey [1].

REMARKS ON TRAINING WORKS.

The training works through estuaries described in this and the preceding chapter, prove that whilst their influence in deepening a channel extends hardly at all beyond their termination, their effects in producing accretion at the sides of an estuary may be apparent a long distance in advance of the trained channel. These results show that, in order to secure a deep stable outlet channel, training works must be carried out to deep water, and that their influence in causing accretion in an estuary must not be lost sight of. The relative advantages of an improved channel, and the maintenance of the tidal capacity of an estuary, must be duly weighed ; and the decision must depend upon the conditions of the case. When the maintenance of the access to a large seaport at the mouth of an estuary depends upon the preservation of the tidal capacity of the estuary above it, no training works should be permitted in the upper estuary ; and therefore training walls in the upper estuary of the Mersey were inadmissible in the interests of Liverpool. For the same reason it would have been inexpedient to carry training works down the Loire estuary, for in benefiting Nantes, St. Nazaire would have been injured. On the other hand, improved outlet channels for the ports in the estuary of the Wash are of more importance than the gradual silting up of the estuary. In the case of the Seine estuary, the claims

[1] ' La Seine Maritime,' P. Mengin-Lecreulx, Vme Congrès International de Navigation intérieure, Paris, 1892, p. 14.

of Havre are paramount; but with this form of estuary, and as Havre is on the open sea-coast, it appears possible, by very carefully designed works, to provide an adequate navigable outlet for Rouen without endangering the approaches to Havre. A more difficult question to decide is how far the improvement of a port up an estuary should override the interests of a seaside town at the outlet, by allowing works to be carried out which would prejudice such towns by producing a progression of the foreshore in front of them, and thus driving the sea further away. Some of the schemes for the prolongation of the Seine training works would produce this effect in front of Trouville (Plate 10, Figs. 1, 2, and 3); and the training walls in the Ribble estuary threaten Southport with a similar fate. Such considerations have hitherto been disregarded, because the wide-spread influence of training walls in a sandy estuary have not been understood; but in future, questions of this nature can hardly be neglected.

Enlargement of Trained Channel. Though the rate of enlargement in width between training walls must depend upon the special conditions of an estuary, such as the tidal rise, fresh-water discharge, slope of river-bed, and size of outlet, some indications of the limits of this enlargement may be obtained from estuaries possessing good outlet channels. The enlargement in untrained estuaries is necessarily irregular; but in those estuaries which possess the most uniform navigable channels, the rate of enlargement increases rapidly on approaching the mouth, and is greater for the high-water channel than for the low-water channel. The increase in width of the Scheldt, between Antwerp and Lillo, is about 1 in 59 in the low-water channel, and 1 in 50 in the high-water channel; whilst between Lillo and Flushing, where both depths and widths vary considerably, it amounts to about 1 in 21 at low water, and 1 in 20 at high water. In the Thames, the rate of enlargement between London Bridge and Gravesend is about 1 in 93 at low water, and 1 in 80 at high water; whilst between Gravesend and Canvey Island it attains about 1 in 30 at low water, and 1 in 8½ at high water; and beyond this latter point both channels widen out rapidly (Plate 7, Fig. 7). In the Humber, the rate of enlargement, from the confluence of the Ouse and the Trent to Sunk Island, is 1 in 48 at low water, and 1 in 37 at high water; whilst from Sunk Island to the mouth at Spurn Point the enlargement is rapid [1] (Plate 7, Fig. 9).

The rate of enlargement adopted for trained channels varies considerably. Thus the widening out of the Seine training walls is only 1 in 200, and of the Vire 1 in 100; whilst the trained channel of the Ribble has been made very nearly uniform between Preston dock and the Naze, though below the Naze to the end of the north training wall the enlargement is about 1 in 80. The enlargement of the trained channel of the Maas, from Schiedam to the mouth, averages about 1 in 77; of the Weser from Bremen to the end of the training walls, 1 in 50; of the Clyde from Renfrew to Dumbarton, 1 in 71; and of the Tyne from Newcastle to its mouth, 1 in 74. The increase in width between the Seine training walls is acknowledged to be much too small; and probably in most cases the suitable rate of enlargement for a trained channel would be found to be comprised between 1 in 80 and 1 in 50, with a more rapid expansion near the outlet. In some cases of training rivers through wide estuaries, too much attention has been devoted to the deepening of the trained channel by scour, and too little regard paid to the maintenance of the tidal capacity of the estuary, leading to an undue narrowing of the width between the training walls; so that in some instances a narrow trained channel emerges abruptly into a wide estuary, as exemplified by the Seine, the Loire, the Ribble, and the Dee (Plate 9, Figs. 1,

---

[1] 'The Physical Conditions affecting Tidal Rivers; and the Principles applicable to their Improvement,' L. F. Vernon-Harcourt, Manchester Congress, 1890, p. 10.

4, 11, and 13). Where, however, dredging is resorted to in combination with training works, a more suitable rate of enlargement can be provided, together with an adequate depth of channel, preventing an excessive reduction of tidal capacity, and abrupt alterations in width and depth at the outlet of the trained channel, as indicated by the training works of the Clyde and the Weser (Plate 8, Fig. 7, and Plate 9, Fig. 8).

Concluding Remarks. It must be always borne in mind, in designing training works through an estuary subject to the introduction of sediment, that no training works can be considered complete unless they guide the channel right out to deep water, and that training walls cause accretion at the sides of an estuary, even in advance of their extremities. Moreover, reliance must not be placed solely upon the scour of the currents for the deepening of a trained channel, since this necessitates an undue restriction of its width, and loss of tidal capacity; but dredging must be employed as an auxiliary to the training works, in order to provide a large enough channel for the free admission of the flood tide. The rate of enlargement of a trained channel seawards, to ensure the free admission of the largest volume of tidal water consistent with the training and improvement of the channel, must depend upon the special physical conditions, the form of the estuary, and the width of its outlet; and it should, as far as practicable, be kept within the limits which nature and experience indicate as the most favourable. The study also of the various states of an estuary, as shown on successive charts, is very valuable in designing the direction for the trained channel, by indicating the action of the natural forces in the estuary, and the position of the most stable and deepest outlet channel. In determining the position of the training walls, provision has to be made for preserving the access to any ports which may exist on the shores of the estuary; and in view of the inevitable accretion resulting from training works, care should be taken in designing the works to avoid as far as possible injuring seaside towns by the accumulation of deposit in front of them.

Reclamation of land is unfavourable to the interests of navigation in an estuary, by reducing the tidal capacity; and though training works have frequently led to reclamation, by stopping the wanderings of the channel and promoting accretion, as in the case of the Ribble, training works should be designed so as to avoid reclamations as far as possible, instead of favouring them, as in the Seine training works, or with the object of reclamation, which proved such a fatal course in the Dee estuary.

CHAPTER XIV.

EXPERIMENTAL INVESTIGATIONS ON TRAINING WORKS IN ESTUARIES.

Origin of Investigations. Advantages of experiments with working Models of Estuaries. Material employed for Bed of Estuary. Arrangement of Models; Method of producing Tidal Flow and Fresh-water Discharge, and Formation of Training Walls. Suitability of Seine Estuary for instituting Investigations on Training Walls. *Experiments with Model of Tidal Seine:* Description of Model; Resemblance of Model to actual Estuary; Changes produced on inserting Training Works; Effects produced on inserting various Schemes for prolonging Training Works, namely, contracted outlet, sinuous channel, and trumpet-shaped channel; Comparison of Results of Schemes. Large Rouen Model described. *Experiments with Model of Mersey Estuary:* Contrast to Seine Estuary; Description of Model; Resemblance of Channels in Model to those in Estuary; Transformation produced by inserting Training Walls in Inner Estuary; Improvement effected by Outer Training Works. Concluding Remarks, reliability of method, its great value for preliminary investigations.

THE great differences of opinion exhibited by engineers in the various schemes proposed for the prolongation of the training walls in the Seine estuary, and the conflicting views expressed as to the influence of training works in the upper estuary of the Mersey, when proposed for the outlet of the Manchester Ship-Canal in 1883 and 1884, led me to endeavour to obtain a definite solution of these disputed problems, which no arguments from analogy appeared capable of effecting, and at the same time to establish the principles of training works in estuaries on a more scientific basis, by experimental investigations with working models of the Seine and Mersey estuaries on a small scale, carried out in 1886-9 The main causes of the uncertainties which exist as to the effects of training works in estuaries, and the divergence of opinion to which schemes of such works give rise, are undoubtedly the very great differences in the configuration and tidal capacity of estuaries, and the varieties of tidal range and fresh-water discharge, so that engineers hesitate to apply the experience gained from training works in one estuary to another estuary subject to more or less different physical conditions. These are the kind of peculiarities which models are specially qualified to reproduce; and in a working model of the inner estuary of the Mersey, constructed in 1885 with the object of showing that the slight regulation of the Cheshire shore of the estuary contemplated in the final design of the Manchester Ship-Canal, now completed (Plate 7, Fig. 11), would not affect the state of the estuary, the existing condition of the inner estuary was very fairly attained [1]. Though the configuration of the shores of an estuary can be exactly imitated in a small scale model, the vertical scale has necessarily to be made much larger than the horizontal scale, in order to produce any appreciable rise of tide and flow of currents; whilst, notwithstanding this exaggeration of the tidal rise, the size of the particles of the finest material that can be procured to represent the shifting sands of an estuary, are very large in a small model in proportion to the actual particles of sand constituting the sandbanks of an estuary. Accordingly, it is not practicable to obtain the same proportionate depths of channels in a small model as in nature; but, on the other hand, a small model can be inspected very closely, and changes in the channels or banks readily noted; and owing to the extreme shortness of the tidal period in a model, alterations which may take years in an estuary, are effected in a short period in a model. The action of waves in an estuary, produced by the prevailing winds, cannot, indeed, be reproduced in a model; but fortunately, in the particular estuaries experimented upon, the direction of the prevalent

[1] 'The Régime of Rivers and Estuaries,' Professor O. Reynolds, Frankfort Congress on Inland Navigation, 1888, p. 11; compare figs. 1 and 2.

winds and of the flood tide entering the estuary approximately coincide, and therefore the tidal impulse given in the model practically represented the combined influences of the flood tide and waves.

Material used for Bed of Model of Estuary. In order that experiments on the effect of training works in an estuary might have any value, it was absolutely essential to produce in the model the accretion which training walls have been found to produce in nature; and the obtaining of this accretion in the model constituted the most novel part of the investigations, and the most difficult effect to reproduce. Various finely divided substances of low specific gravity were successively tried, but the fine particles became gradually consolidated into too compact a mass to be readily affected by the minute currents; whilst the sawdust of heavy woods became swollen in the water, and was too easily shifted by the currents[1]. The only material tried which at all satisfied the requisite conditions, was very fine sand containing peat, procured from the Bagshot beds on Chobham Common. This sand, being perfectly clean, did not become a compact mass; and as the water was readily able to percolate through it, the finest particles could be transported by the currents.

Arrangement of Models for Investigations. The first step required in an investigation of this kind is to reproduce in a model a previous, or existing condition of the estuary to be investigated. For this purpose, in the two estuaries experimented on, the estuary, with the whole of the tidal portion of the river above it, was moulded in Portland cement to its exact form, though to a distorted scale vertically; and the model was also extended to an adequate distance beyond the mouth of the estuary. The shifting portions of the bed of the estuary were formed with fine Bagshot

---

[1] 'The Principles of Training Rivers through Tidal Estuaries, as illustrated by Investigations into the methods of improving the Navigation Channels of the Estuary of the Seine,' L. F. Vernon-Harcourt, Proceedings of the Royal Society, vol. 45, p. 511.

sand, laid to an adequate depth over the cement bottom ; and the fresh-water discharge was let in through a small tap from a cistern at the tidal limit. The tidal flow and ebb were effected by the raising and lowering of a zinc tray of suitable size, enclosed on three sides, and connected with the model on the open side by an indiarubber hinge. The tray was fastened to the model at an angle so adjusted that the water flowing into the model estuary, on lifting the tray, entered in the direction that the flood tide actually approaches the estuary. The size of the tray also, and the amount its outer end was raised, were arranged so that the rise of the water in the model corresponded with the rise of tide at the mouth of the estuary, according to the small vertical scale adopted. The time occupied in each raising and lowering of the tray, or the tidal period of the model, was adjusted so as to reproduce in the. small-scale model the tidal phenomena of the river under examination, such, for instance, as the time required by the tide to reach the end of the tidal portion of the river in relation to the tidal period.

As soon as it was found that, after making suitable adjustments, the working model fairly represented in miniature the actual condition of the untrained estuary, with its shifting channels and other peculiarities, strips of tin representing the training walls to the distorted vertical scale, bent to the desired lines, were introduced; and the gradual changes which resulted from their introduction, after working the model for a great number of tides, were carefully recorded on a chart of the estuary.

Suitability of Seine Estuary for Primary Investigations. The correspondence obtained between the state of the estuary in a model and the natural condition of the estuary, might afford a presumption that the effects of training works in a model would represent approximately the effects such works would produce in the actual estuary. Owing, however,

to the influence exercised by accretion on the changes in
an estuary produced by training walls, it would be impossible
to prove, by the preliminary step in the investigations, that
the training walls in the model would be able to exercise
the same influences with regard to accretion as corresponding
works in the real estuary. In fact. the crucial question in the
investigations was, whether accretion could possibly be made
to result from the introduction of training walls in the model,
with the comparatively coarse material composing the bed of
the miniature estuary, in at all a similar manner as had been
witnessed in various estuaries. Fortunately the Seine estuary
was specially suitable for the application of such a test to
a model, for its state previously to the commencement of the
training works is recorded in early charts, as well as its
condition since the completion of the works, by various
surveys. If, besides the first step of making a model fairly
representing the original condition of the Seine estuary
previously to the works, it was possible to reproduce in the
model the actual changes produced by the training works, it
might fairly be inferred that the model sufficiently accorded
with the condition of the estuary to indicate future changes
from any prolongations of these works. On the accomplish-
ment of this second step, the model becomes a most valuable
guide as to the effects, and relative advantages, of the various
conflicting schemes proposed for the completion of the
training walls in the estuary of the Seine; and the system
can be applied with confidence to ascertain the effects of
training works in any estuary with shifting sandbanks.

EXPERIMENTS WITH MODEL OF TIDAL SEINE [1].

The remarkable variety in the schemes proposed for the
completion of the Seine training works was referred to in

---

[1] 'The Principles of Training Rivers through Tidal Estuaries, as illustrated by
Investigations into the Methods of improving the Navigation Channels of the
Estuary of the Seine,' L. F. Vernon-Harcourt, Proceedings of the Royal Society,
vol. 45, pp. 504–524, and plates 2, 3, and 4.

the last chapter; and the chief types are given in Plate 10, Figs. 1 to 6, though several others have been put forward in various publications [1]. The frequent appearance of fresh schemes, and the conflict of views thereby exhibited, led to my undertaking these investigations; and the Seine estuary not only provided ample material for experiments with these divergent schemes, but also afforded the means of determining whether it was possible to reproduce with sufficient exactness the effects of training works in a model.

Description of Tidal Seine Model. The model of the tidal Seine, made to the scales of $\frac{1}{10000}$ horizontal and $\frac{1}{400}$ vertical, extended from the tidal limit at Martot down to about Dives on the Calvados coast, to the south-west of Trouville. The tidal rise at springs, produced by the extreme tipping of the tray, amounted to 0.71 inch in the model; and the tidal period was about 25 seconds. The bed of the estuary was formed of a layer of fine Bagshot sand, except where rock crops up to the surface in front of Villerville, and the solid portions of the Amfard and Ratier banks at the outlet consisting of gravel or rock, which were moulded in Portland cement. The hinge of the tray was arranged so that the tidal influx took place from a direction about 5° to the west of north-west. The fresh-water discharge, proportionate to the scale of the model, was admitted at the upper end from a tank; and a corresponding quantity was discharged at the outer end of the model during the higher half of the tide, to prevent an increase in the volume of water in the model.

Results obtained on working the Seine Model. From the commencement, the bore was well marked by a sudden rise, particularly at the point of the model representing Caudebec; and the reverse current, which flows out of the estuary near Havre just before high water, was apparent, resulting from the filling of the estuary by the second tide following the southern shore, which was produced by the angle at which the tray

[1] Minutes of Proceedings Institution C. E., vol. lxxxiv, plate 4, fig. 9.

was placed to the estuary and the prolongation of the southern coast. Even with the silver sand first used, shifting channels were formed ; and a condition resembling fairly closely a previous state of the estuary was obtained after the model had been worked for some time. The silver sand, however, proved too coarse to be readily moved by the feeble currents of the model, and failed to produce accretion when the training walls were inserted. Various other fine substances also did not adequately fulfil the requisite conditions, or reproduce satisfactorily in miniature the state of the Seine estuary after the introduction of the training works.

Insertion of Seine Training Works in Model. At last, with a bed of Bagshot sand, the condition of the estuary, as indicated in the chart of 1834, was obtained in the model ; and accretion was produced on the insertion of the training walls. Strips of tin were fixed in the model to the heights, and in the positions corresponding to the actual works, and as far as practicable in a similar sequence, so as to conform as nearly as possible to the same conditions in the model as in the estuary. The training walls in the model produced a deepening in the trained channel, and an accretion at the sides of the estuary behind the walls, closely resembling the actual changes in the estuary ; and, moreover, the accretion extended down to Hoc Point on the northern side, obliterating the Harfleur channel, and down to Honfleur on the southern shore, corresponding precisely in extent to the accretion which has taken place in the estuary.

Effects of various Schemes on Model of Seine Estuary. The success at last obtained in the second stage of the investigations, enabled the third, and final stage to be entered upon with confidence. As it had been proved that the model, with a bed of Bagshot sand, could reproduce the results of training works actually carried out, the effects of prolongations of these works in the model, according to the different schemes, might reasonably be regarded as foreshadowing the effects which these works would actually produce if carried out in the estuary. Accordingly, some of the typical schemes were inserted successively in the estuary; and the model was worked for from 3500 to 7300 tides for each scheme, with the results shown on Plate 10, Figs. 1 to 6.

The contracted outlet formed by a breakwater extending from the southern shore at Villerville to the Amfard bank, as proposed in some of the schemes, converting the estuary into a large sluicing basin, naturally produced a deep outlet channel between the Amfard bank and Havre in the model (Plate 10, Figs. 1 and 2). The inside channel, however, was tortuous in the first case, and shallow above Honfleur in the second arrangement, owing to the divergence in the directions of the flood and ebb currents, and the gradual silting up of the estuary in the model, from the deposit resulting from the loss of velocity experienced by the rapid silt-bearing flood-tide current through the narrow outlet on expanding in the sheltered estuary. Moreover, the contraction at the mouth appreciably reduced the tidal rise inside, notwithstanding the minute scale of the model; whilst a bar appeared between the deepened outlet channel and deep water in the sea. The breakwater also, by checking the silt-bearing flood tide along the southern coast, and protecting this shore from the out-flowing ebb, caused a great advance of the foreshore outside it in front of Trouville in the model; and by the shelter it afforded on its inner side to the southern shore of the estuary, it occasioned a similar progression of the foreshore between Villerville and Honfleur in the model. The accretion in the model estuary gradually reduced the great initial depth of the narrow outlet channel between Amfard and Havre, by diminishing the tidal capacity of the scouring basin which formed it.

The results of working the model with the prolongation of the trained channel in a sinuous form, as advocated in some proposals (Plate 10, Figs. 4 and 5), indicate clearly the disadvantages of the want of coincidence in the direct course of the flood tide and the winding course of the ebb, occasioned by this form of channel. In neither of the cases investigated was a deep, direct, or continuous channel obtained by this form of training works in the model.

The schemes providing a trumpet-shaped outlet channel furnished the most satisfactory results in the model (Plate 10, Figs. 3 and 6), giving wider and more uniform channels than the other types of schemes. The arrangement in which the channel was given the least expansion, and a continuously concave southern training wall to the outlet, afforded a some-what more uniform channel in the model than the other; but, at the same time, the narrower outlet occasioned a greater accretion inside the estuary, and a considerable progression of the foreshore in front of Trouville in the model. Whilst a restriction of width between the training walls produces a rather more direct channel, the maintenance of the width at the outlet by a greater expansion of the training walls preserves a larger tidal capacity, and obviates great changes in the estuary and near the outlet. On the whole, a more expanded channel, in which deficiencies in depth or direction could be remedied by dredging, or possibly by low inner training works, appears preferable to a narrower trained channel deepened entirely by scour, on account of the greater loss of tidal capacity, and more important changes in the natural conditions produced by a considerable reduction in the width of the outlet.

Comparison of the Schemes investigated. The conversion of the Seine estuary into a vast scouring basin by a break-water greatly contracting the outlet, though securing deep water at the mouth, evidently would produce too extensive modifications in the estuary and adjacent foreshore to be admissible. Moreover, the irregular depth of the channel thereby formed, the variation in the velocity of the currents at different parts, and the great eventual loss of tidal capacity in the estuary, due to inevitable accretion, producing a great reduction in the tidal influx and efflux at the outlet, would preclude the adoption of any scheme of this nature.

The sinuous type of trained channel, by ensuring a divergence between the flood and ebb currents, and a consequently unsatisfactory channel, should evidently be avoided, as far as practicable, in tidal rivers. Accordingly, schemes of this form are inexpedient for the training of rivers through tidal estuaries, being based upon the stability of the channel of a non-tidal river at curves, which does not hold good when the current is alternately in opposite directions.

The only schemes for prolonging the Seine training walls which gave satisfactory results in the model were those which expanded to a trumpet-shaped outlet. Before any definite scheme is adopted, the most suitable form and rate of expansion should be determined by further experiments; but by resorting to dredging, aided possibly by low-water inner training works, it should be possible to adopt a sufficiently expanded channel to avoid the occurrence of serious changes in the condition of the estuary and adjacent foreshores.

Rouen Model of Tidal Seine. In consequence of the results of these experimental investigations, and in accordance with the views expressed by me after giving a brief account of my experiments at the Frankfort Congress in 1888[1], the government engineers at Rouen have made a large model of the tidal Seine, with the object of further investigating the proposed schemes. At first a reservoir was adopted, in place of the winding upper channel of the tidal river; but it proved necessary to complete the model up to the tidal limit, as done at the outset in the earlier model. The Rouen model now looks exactly like an enlarged copy of the little model; and the same angle was adopted for the introduction of the flood tide. The rise of tide, however, is effected by the immersion

---

[1] Verhandlungen der Allgemeinen und Abtheilungs-Sitzungen, III Internationaler Binnenschifffahrts-Congress zu Frankfurt am Main, 1888, p. 193.

of blocks in little wells, worked by a gas engine, instead of the tipping of a tray. The horizontal scale of the model is $\frac{1}{5000}$; and allowance has been made for using a vertical scale of $\frac{1}{100}$ or $\frac{1}{50}$, with a tidal period of 60 or 100 seconds. A small maregraph enables the miniature tidal curves to be registered at certain points on the model.

The comparatively large size of the model should enable very accurate investigations of the relative effects of the various schemes to be carried out; and the results obtained will be of the highest interest, both in determining the general form that should be adopted for the completion of the training works in the Seine estuary, and also in indicating the value of these experimental investigations for the solution of the difficult and intricate problems involved in the training of rivers through tidal estuaries. In working the Rouen model, it appears that the preliminary steps of the investigations, of reproducing a former state of the estuary, and then effecting the changes produced by the existing training works, were dispensed with, and that the influences of some of the schemes for the prolongation of the training walls, on the model estuary, were investigated in the first instance. These preliminary experiments seem always expedient; and the substitution of fine Bagshot sand for the coarser sand hitherto used in the model, will be advantageous.

EXPERIMENTS WITH MODEL OF MERSEY ESTUARY [1].

The Mersey estuary, with a wide inner estuary, followed by a narrow neck opening abruptly into the wide expanse of Liverpool Bay, and with a comparatively small amount of fresh water flowing into it from the Mersey and the Weaver, offers a striking contrast in form and condition to the Seine

[1] Proceedings of the Royal Society of London, vol. 47, p. 142; 'Effects of Training Walls in an Estuary like the Mersey,' L. F. Vernon-Harcourt, London, 1890; and 'Amélioration de la Partie Maritime des Fleuves y compris leurs Embouchures,' L. F. Vernon-Harcourt, Vme Congrès International de Navigation intérieure, Paris, 1892, p. 27.

estuary (compare Plate 7, Fig. 11, with Plate 9, Fig. 1). It, accordingly, afforded an excellent opportunity for the extension of the system of experimenting on training works, by means of models, to a totally different estuary. As no training works have been carried out in the Mersey estuary, the second stage of the Seine experiments could not be effected ; but the satisfactory results obtained in the Seine model dispensed with the necessity of a similar stage in the case of the Mersey model, in which similar arrangements, and a similar material for the bed of the estuary were adopted.

Description of Mersey Model. The model of the Mersey extended from Warrington to beyond the bar in Liverpool Bay, and was made to a horizontal scale of $\frac{1}{30000}$, and a vertical scale of $\frac{1}{500}$. The exaggeration of the vertical scale was made less than in the Seine model, on account of the greater rise of tide at the mouth of the Mersey, and the greater fall of the bed of the estuary, than in the case of the Seine. The shore-lines of the model were extended to Formby Point on the northern side, and to the sandbank separating the Mersey estuary from the estuary of the Dee on the southern side; and the sides of the model were prolonged in slightly diverging lines seawards up to the hinge of the tray, which was placed a little distance beyond the site of the bar. The tray was so adjusted that the flood tide was introduced into the model from a direction 17° 15′ to the north of west, which appears to be approximately the direction from which the tide enters Liverpool Bay. The tray, on being raised to its full extent, produced a rise of $\frac{3}{4}$ inch opposite the site of Liverpool, which corresponds to the equinoctial spring-tide rise of 31 feet on the scale of the model; and the tidal period was 32 seconds in the model. The proportionate small flows of water representing the fresh-water discharges of the Mersey and Weaver, were let in from two small cisterns at Warrington and Frodsham respectively in the model.

Results of working Mersey Model. The model was first worked in its natural condition, to ascertain how nearly the existing condition of the Mersey estuary could be reproduced in the model. In working the model for a number of tides, irregular channels appeared in the bed of Bagshot sand in the inner estuary, which constantly shifted; and two more permanent channels skirted each shore above the 'narrows,' up to Eastham and Garston respectively in the model (Plate 10, Fig. 7). A deep channel was formed in the 'narrows' between Liverpool and Birkenhead in the model, and extended into the bay with a gradual reduction in depth out to the bar, which appeared in the model approximately in its actual site, showing a tendency to shift its position slowly, like the Mersey bar. A channel was also formed along the Cheshire shore in the bay, at right angles to the main channel, corresponding to the well-known Rock Channel; and a shallow subsidiary channel diverged from the main channel towards Formby Point, which exists in the actual estuary (compare Plate 7, Fig. 11, with Plate 10, Fig. 7). On the whole, considering that it is impossible to produce the action of breaking waves in a model, which undoubtedly causes the rugged appearance of the sandbanks in Liverpool Bay, the model reproduced remarkably closely the main features of the Mersey estuary, confirming, in the case of a very different estuary, the practicability of exhibiting in a small model the general conditions of an estuary out to the open sea, previously established with regard to the Seine.

Experiment with Training Walls in Inner Mersey Estuary. Two problems of considerable interest present themselves for investigation with regard to an estuary of the peculiar form of the Mersey, namely, the effect of training works in the inner estuary on the general condition of the estuary, and the influence that training works in the outer estuary might have on the outlet channel.

The proposals made in 1883 and 1884, to provide an outlet

for the Manchester Ship-Canal from Runcorn to deep water, by training works through the inner estuary of the Mersey, gave rise to most divergent opinions amongst engineers. It was admitted that the maintenance of the depth in the outlet channel below Liverpool, and over the bar, depended on the preservation of the tidal capacity of the inner estuary. The difference of opinion arose on the question whether the formation of a trained channel for the river through the inner estuary would, or would not, give rise to accretion, and consequently affect the tidal capacity of this part of the estuary. Though this scheme was abandoned on its rejection by a Select Committee of the House of Commons in 1884, after passing through the House of Lords, and the present line of canal, skirting the Cheshire shore down to Eastham, was consequently substituted, the effect of training works under such conditions is of considerable scientific interest ; and it is not advisable that this subject should remain in doubt, as a similar problem might at any time arise in some other estuary. Accordingly, strips of tin were inserted in the Mersey model, representing the training walls proposed for the outlet channel of the canal through the upper estuary. After working the model for some time with this modification, a remarkable change had become apparent in the estuary. The channel between the training walls had been scoured out, but the rest of the upper estuary had been filled up ; and the consequent reduction in tidal capacity had produced a great deterioration in the outlet channel beyond the 'narrows,' thus fully confirming a view expressed in 1881, before the canal scheme was contemplated [1], and the fears entertained by the

[1] ' The condition of the river above Liverpool has formed the subject of considerable discussion. It would be impossible to improve its depth without confining its channel by training banks, which would lead eventually to the reclamation of a large portion of the existing tide-covered area between Runcorn and Liverpool. The gain to the river above Liverpool from such works would be considerable ; but the reduction of the tidal receptacle would reduce the scouring power of the current in the estuary below, and would in all probability lead to a deterioration of the outlet.' ' Rivers and Canals,' 1st edition, p. 257.

Liverpool authorities on the introduction of the project (Plate 10, Fig. 8). The result of this experiment proves conclusively that training works should never be put into a portion of an estuary subject to the introduction of sediment, whenever the maintenance of its tidal capacity is essential for the preservation of the depth of the outlet channel of a large port situated lower down on the estuary..

Experiments with Training Walls in Outer Mersey Estuary. The improvement in the depth over the Mersey bar, which is being effected by dredging, has been previously referred to (p. 279). The only other way in which the depth over the bar might be increased, is by regulating the outlet channel beyond New Brighton by training works, which would render the channel more stable, and produce a greater scour across the bar. Training works would be costly in such an exposed situation, especially if they had to be carried some distance out; but they would obviate the constant maintenance necessitated in deepening the channel by dredging alone. The question, however, possesses a general interest irrespectively of its special application to the Mersey; and it was mainly as a novel problem that the experiments with training walls in the outer estuary of the Mersey model were undertaken. The training walls inserted in the model in this instance, were designed to form a trumpet-shaped outlet in continuation of the shore-lines of the 'narrows,' so as to promote, rather than impede, the influx of the tide into the 'narrows' and inner estuary, whilst restricting the ebb and flow to a single channel (Plate 10, Fig. 9). On working the model with this arrangement, the outlet channel became fixed and deeper; whilst the maintenance of the inner estuary was not affected. The general direction and depth of the outlet channel was improved by removing the northern portion of the southern bank, which protruded somewhat into the trained channel, showing that dredging is a very useful auxiliary to training works. It appears probable from the experiments that in the actual estuary, the outlet channel might be regulated and deepened by a single training wall extending seawards for a moderate distance from New Brighton, diverging somewhat from the Lancashire shore, and shutting off the Rock Channel; and this arrangement, besides rendering the channel over the bar more stable, would reduce the amount of dredging required for maintenance.

The experiments with the Mersey model show that in a very irregular and defective estuary, training works may be injurious in the upper portion, and advantageous below where accretion at the sides does not affect the maintenance of the outlet channel.

Concluding Remarks. The experiments carried out with the models of the Seine and the Mersey show that definite, and apparently reliable, indications can be obtained in this manner of the effects of training works in any tidal estuary, since definite and concordant results were obtained from models of two such very dissimilar estuaries as the Seine and the Mersey. Considering the unexpected results that have followed training works in estuaries, the divergence of opinion on the subject exhibited by engineers, and the facility with which the influence of various schemes may be tested in a working model, it may be hoped that in future no training works will be carried out in estuaries without seeking the guidance so readily afforded by the experimental method. By this means, injurious works leading to irremediable consequences may be avoided, and the best out of several alternative schemes may be selected. The system of experimenting with models of estuaries promises to prove of inestimable value to engineers in the design of training works through sandy estuaries; and each set of such experiments will aid in the establishment of definite principles with reference to this complex branch of the science of river engineering.

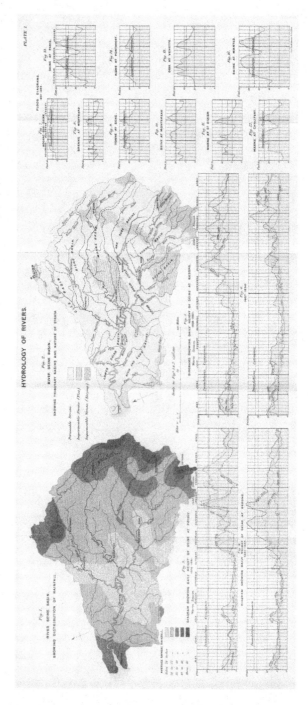

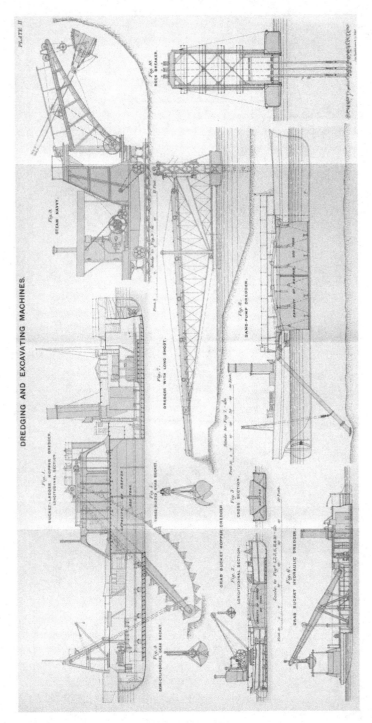

PLATE II

DREDGING AND EXCAVATING MACHINES.

Fig. 1.
BUCKET-LADDER HOPPER DREDGER.
LONGITUDINAL SECTION.

Fig. 2.
GRAB BUCKET HOPPER DREDGER.
LONGITUDINAL SECTION.

Fig. 3.
CROSS SECTION.

Fig. 4.
SEMI-CYLINDRICAL GRAB BUCKET.

Fig. 5.
THREE-BLADED GRAB BUCKET.

Fig. 6.
GRAB BUCKET HYDRAULIC DREDGER.

Fig. 7.
DREDGER WITH LONG SHOOT.

Fig. 8.
SAND-PUMP DREDGER.

Fig. 9.
STEAM NAVVY.

Fig. 10.
ROCK BREAKER.

The material originally positioned here is too large for reproduction in this reissue. A PDF can be downloaded from the web address given on page iv of this book, by clicking on 'Resources Available'.

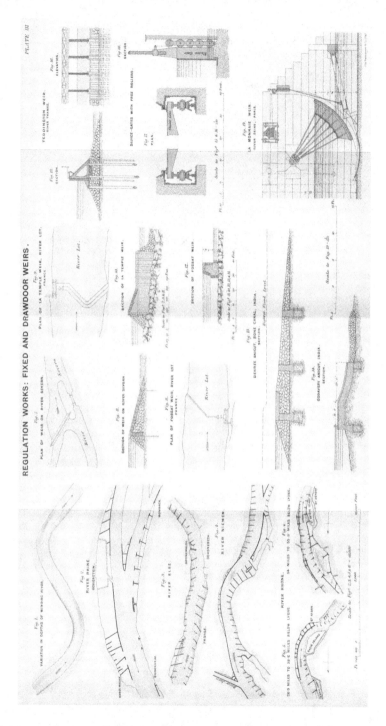

PLATE III

REGULATION WORKS: FIXED AND DRAWDOOR WEIRS.

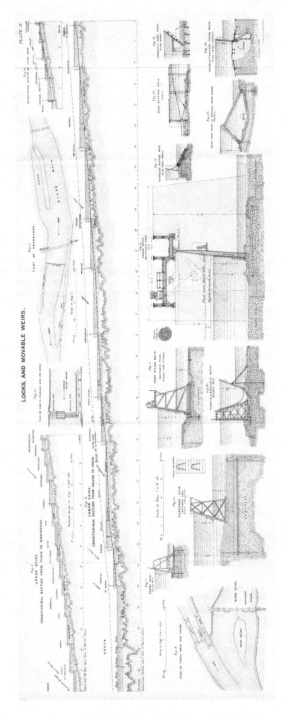

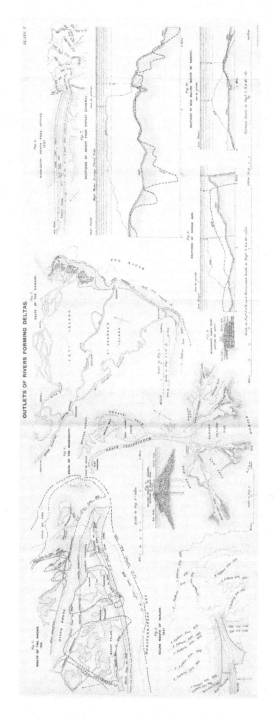

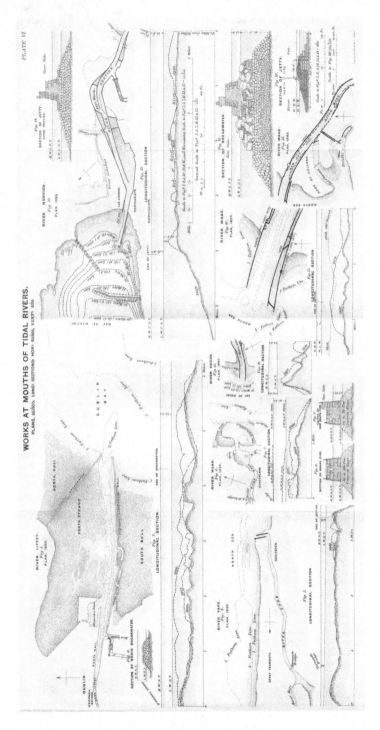

WORKS AT MOUTHS OF TIDAL RIVERS.

PLANS, ½₁₅₀₀. LONG⁰ SECTIONS HORI. ½₄₂₀₀₀, VERT⁰ ⅙₆₀.

PLATE VI

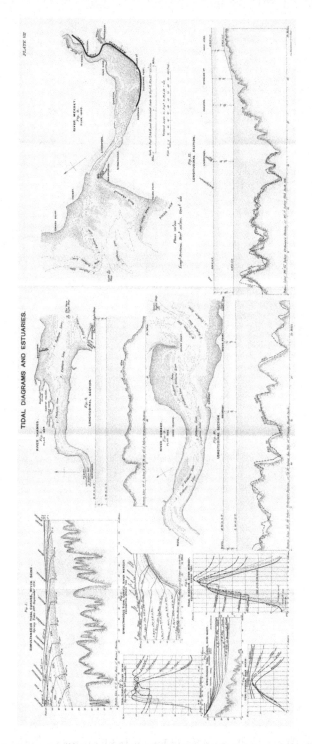

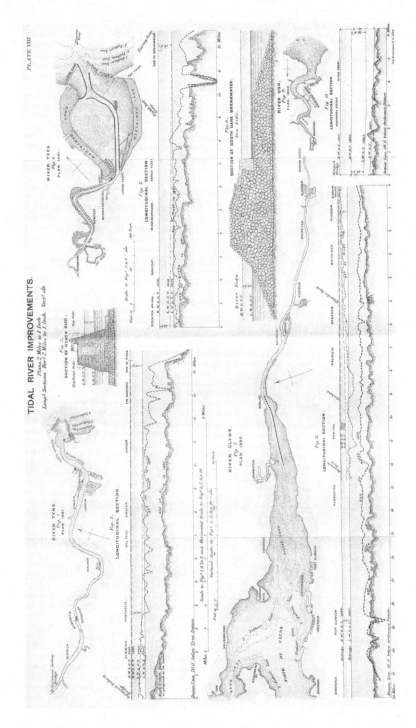

PLATE VIII

TIDAL RIVER IMPROVEMENTS.
Plans. 2 Miles to 1 Inch.
Longl Sections. Horl 2 Miles to 1 Inch. Vertl &c

The material originally positioned here is too large for reproduction in this reissue. A PDF can be downloaded from the web address given on page iv of this book, by clicking on 'Resources Available'.

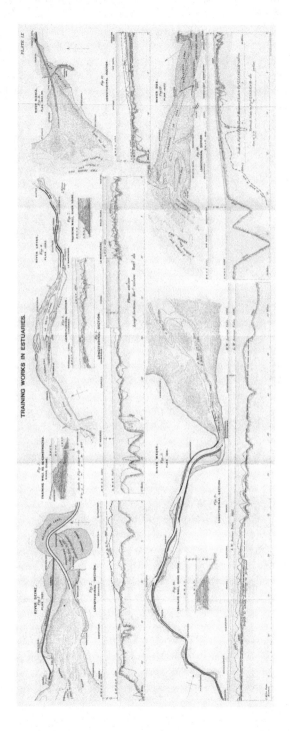

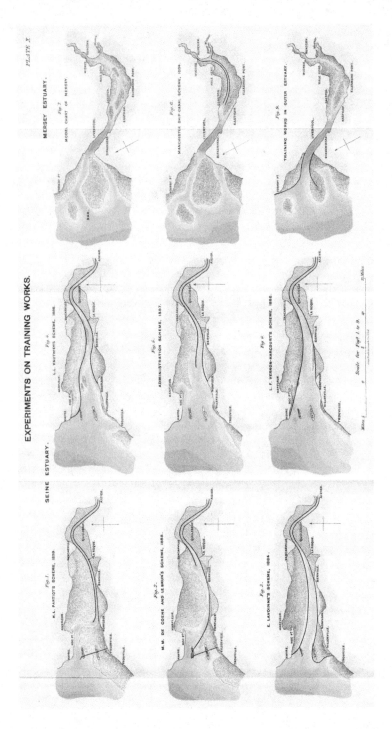

Printed in the United States
By Bookmasters